# SOLVING COMPLEX PROBLEMS FOR STRUCTURES AND BRIDGES USING ABAQUS FINITE ELEMENT PACKAGE

# SOLVING COMPLEX PROBLEMS FOR STRUCTURES AND BRIDGES USING ABAQUS FINITE ELEMENT PACKAGE

Farzad Hejazi and
Hojjat Mohammadi Esfahani

**CRC Press**
Taylor & Francis Group
Boca Raton London New York

CRC Press is an imprint of the
Taylor & Francis Group, an **informa** business

First edition published 2022
by CRC Press
6000 Broken Sound Parkway NW, Suite 300, Boca Raton, FL 33487-2742

and by CRC Press
2 Park Square, Milton Park, Abingdon, Oxon, OX14 4RN

© 2022 Farzad Hejazi and Hojjat Mohammadi Esfahani

CRC Press is an imprint of Taylor & Francis Group, LLC

*Library of Congress Cataloging-in-Publication Data*
Names: Hejazi, Farzad, author. I Esfahani, Hojjat Mohammadi, author.Title: Solving complex problems for structures and bridges using ABAQUS finite element package / Farzad Hejazi, Hojjat Mohammadi Esfahani.
Description: First edition. I Boca Raton, FL : CRC Press, 2022. I Includes index.
Identifiers: LCCN 2021025807 (print) I LCCN 2021025808 (ebook) I ISBN 9781032100395 (hbk) I ISBN 9781032100401 (pbk) I ISBN 9781003213369 (ebk)
Subjects: LCSH: Structural engineering--Mathematics. I Bridges--Design and construction--Mathematics. I Finite element method--Data processing. I Abaqus (Electronic resource)
Classification: LCC TA647 .H45 2022 (print) I LCC TA647 (ebook) I DDC 624.1/71--dc23
LC record available at https://lccn.loc.gov/2021025807
LC ebook record available at https://lccn.loc.gov/2021025808

ISBN: 978-1-032-10039-5 (hbk)
ISBN: 978-1-032-10040-1 (pbk)
ISBN: 978-1-003-21336-9 (ebk)

DOI: 10.1201/9781003213369

Typeset in Times
by MPS Limited, Dehradun

# Contents

# Preface

Finite element is a numerical method that can be employed in solving a complex problem in which no classical solutions can be applied. It is based on the usage of matrix and is carried out with the help of computers. The finite element was developed half a century ago and has since been used in various fields of engineering and applied sciences. It can also be used to obtain an approximate solution for engineering analysis.

The Abaqus software is one successful finite element package which was developed in 1978 and has been used since then for numerical simulation of complex problems in various fields, such as Civil and especially Structural Engineering. The Abaqus is a general-purpose finite element simulation package used for mainly numerically solving a wide variety of design engineering problems. Many of these problems include static/dynamic structural analysis (both linear and nonlinear), heat transfer and fluid problems, as well as acoustic and electromagnetic problems. Abaqus contains many special tools that assist in analyzing various effects such as plasticity, large deformations, hyper-plasticity, creep, high deflection, interface contact, the dependence of various properties on temperature and pressure, anisotropy, and radiance. The possibilities of substructures, submodeling, and optimization can also be determined by adding subroutines to the application. Therefore, the use of computer software such as Abaqus is vital and necessary to analyze and solve engineering problems.

Although implementing Abaqus for solving many issues in various fields of Civil Engineering has been considered by many experts, however, its application to address complex problems in Structural Engineering is highly complicated. Therefore, this book aims to present specific complicated and puzzling challenges encountered during application of the finite element method in solving Structural Engineering problems by using Abaqus software, which can fully utilize this method in complex simulation and analysis.

Therefore, an attempt has been to demonstrate all the processes of modeling and analysis of impenetrable problems through simplified step-by-step illustration with presenting screenshots from software in each part and also showing graphs.

After reading this book, the skill and knowledge of the readers pertaining to modeling and analysis of complicated and convoluted problems in Structural Engineering will be incredibly enhanced and they will be able to:

- Solve general and complicated problems in the Structural Engineering field by using Abaqus software in a reasonable timeframe and with reasonable efforts.
- Develop subroutine for Abaqus software to solve the special problems.
- Gain the ability in analyzing and modeling different types of structures with various conditions.
- Easily use the Abaqus software to simulate irregular-shaped objects composed of several different materials with multipart boundary conditions.

- To complete various tutorials from a broad array of applications such as bridges, offshores, dam, seismic resistant systems, and so forth.
- Implement the available tools in Abaqus software such as scripting mesh and writing subroutines.
- Apply various load effects such as dynamic, explosive loading, seismic excitation, long-term fluid influx, and impact load to the developed structural models in Abaqus software.
- Implement various types of new materials such as Ultra High-Performance Fiber Concrete in special structures such as high-rises or towers by considering damage plasticity models.
- Simulate the fluids penetration in the porous medium and soil consolidations.
- Extract the modal and buckling modes to estimate structures tolerance limits.
- Implement the optimization tools to define design constrains, design objectives and restrictions for the best available conditions.

# Authors

**Farzad Hejazi** is an Associate Professor at the Department of Civil Engineering, Faculty of Engineering, University Putra Malaysia (UPM), and a Senior Visiting Academic at the University of Sheffield. He is an innovation champion in UPM since 2013, and a member of the management committee for the Housing Research Center in Faculty of Engineering, UPM. He was appointed as the Innovation Coordinator for Faculty of Engineering in 2014 by the deputy vice chancellor for Research and Innovation, and was also appointed Research Coordinator for Department of Civil Engineering in 2017. He received his PhD in Structural Engineering from the University Putra Malaysia in 2011 and worked as a postdoctoral fellow until 2012 and thereafter employed as a member of department of Civil Engineering, UPM.

**Hojjat Mohammadi Esfahani** received his Bachelor of Mechanical Engineering in 2007 and his Master of Science in the same field of Mechanical Engineering in 2017. His main expertise is Finite Element Simulation, and he has more than 10 years of experience in teaching and training Finite Element Packages such as Abaqus. Currently he is involved with many research and industry projects regarding simulation of complex and infrastructures by using Abaqus Finite Element Software.

# 1 Introduction about Finite Element Method

## 1.1 INTRODUCTION

Analysis of engineering problems using numerical methods is result of solving a number of coupled and uncoupled equations. These equations are derived based on the laws of physics and mathematics, including equilibrium forces, energy stability, mass survival, thermodynamics laws, Maxwell theory, and Newtonian laws of motion. However, the mathematical models based on these equations are mostly complex and impossible to solve manually, except for very regular geometries, such as circles and rectangles with simple boundaries. Therefore, the use of numerical methods in solving these coupled and uncoupled equations is vital.

The finite element method (FEM) is a part of various numerical solution processes, which is widely used in most of the fields of engineering, such as solid mechanics, structures, fluids, heat transfer, sound, etc. Therefore, nowadays the FEM is being considered an essential aspect in analysis and design in a majority of the engineering fields. This method is undoubtedly one of the most powerful and versatile tools for solving basic engineering problems. However, as FEM requires simultaneous analysis procedures, it may be considered as a very complicated tool by researchers, engineers, and students.

As with many scientific phenomena, the exact history of FEM is unknown. However, evidences provide adequate proof that Richard Courant used this method to analyze complex structures in civil and aerospace engineering in 1943. At that time, no specific name was given to this method, and the method which Courant used involved breakdown of continuous medium (material) into a series of smaller pieces of material called elements (components).

As mentioned before, the FEM is widely being implemented in solving and analyzing of various engineering problems and it is expected to make a significant progress in the near future. Due to rapid advancements in personal computers, supercomputers, and other computation systems that help solve engineering problems, the development of FEM and its practical applications have increased significantly over the recent decades.

## 1.2 MAIN PROCESS FOR FINITE ELEMENT MODELING AND ANALYSIS

The main process for finite element modeling and analysis are summarized in seven following steps:

DOI: 10.1201/9781003213369-1

## 1.2.1   STEP 1: IDEALIZATION

In this step, based on the considered problem and its geometry and conditions, the problem is idealized as 1D, 2D, or 3D problem or axisymmetric, plane stress, or plane strain problem. The computational volume of the analysis is directly related to the number of degrees of freedom of the model. The number of degrees of freedom located on each node can be determined according to the problem definition in 2D or 3D and the type of analysis. For example, for stress–strain analysis in 3D models, six degrees of freedom (translation in three directions $x$, $y$, and $z$ and rotation around each of these three axes) and in 2D models, three degrees of freedom (translation in two directions of $x$ and $y$ and the rotation around the $z$-axis) is allocated for each node.

## 1.2.2   STEP 2: DISCRETIZATION

In this step the considered geometry discretizes into discrete components called elements. The elements are connected to each other using points which are called nodes. Dividing and shredding the geometric shape is the most crucial process in the FEM.

Finite elements and nodes form the geometry of analytical models. Elements are small geometric shapes of a member, which contain specific and complete properties of an analytical model and are connected by nodes. In fact, the number and location of nodes define the shape of the elements. The location of the nodes and the connection of elements to each other determine the final coordinates and dimensions of the member geometry.

### 1.2.2.1   Meshing of Model

The set of all elements and nodes is called a mesh. The mesh and the main geometry of the model are usually overlapped and compatible. Depending upon the shape of the member, the mesh size can be determined as fine mesh or coarse mesh. Although using fine mesh leads to accurate results, however, it is utilized in huge computational processes that are time-consuming and cost a lot. Therefore, the best mesh size is the optimum element size to give accurate results with appropriate computational process. An example of discretization and meshing of various parts for a problem regarding soil–structure interaction is showed in Figure 1.1.

**FIGURE 1.1**   Discretization and meshing of various parts of considered geometry.

In addition to the nodes and elements, other components such as integration points and also cross sections play an important role in meshing the model to obtain accurate results. Although these features are not visible in the finite element package, however, the user can define the corresponding parameters, which will be further demonstrated in future chapters.

The higher density of the elements usually enhances the accuracy of the results where they converge to a single value; thereby, increasing the solving time required by the processor. It should be noted that simply increasing the compaction density does not lead to accurate results, but the correct modeling considerations and assumptions are required. This point is often overlooked by many novice users.

### 1.2.3 STEP 3: ELEMENT CHARACTERISTICS

In the third step, the characteristics of elements such as material properties, stiffness, and damping (for dynamic analysis) are required to be defined. The stiffness of each type of elements can be derived through constitutive and analytical model of corresponding element, which is dependent on shape of the element and the degree of freedom of the element's node. Also, accurate parameters are required to define a material, which may be time-consuming and also require a variety of laboratory tests. But it should be noted that the validity of the FEM analysis results is directly related to the accuracy of the material's parameters. Many special materials (such as special steels, bronze, wood, etc.) are not included in the predefined library of many finite element packages, and it is required to define these material details by the users.

### 1.2.4 STEP 4: ASSEMBLY OF FINITE ELEMENT EQUATIONS

Once the characteristics of all individual elements are defined in terms of matrix, such as stiffness matrix, mass matrix, and damping matrix, then the overall characteristic of considered model geometry is obtained through assembly of individual characteristic matrix for elements according to assigned degree of freedom for nodes of each element. Thereafter, the finite element equations are derived for static or dynamic analysis by defining and adding force vector to the assembled matrixes.

### 1.2.5 STEP 5: APPLY BOUNDARY CONDITION

In the further step, the boundary conditions affecting the solution process must be determined. These conditions include loading protocol, boundary conditions, initial conditions, interactions, constraints, and connections. In general, any phenomenon that causes a stimulation is called "loading," which can be defined as force (at point, edge, surface, or volume), heat flux, electrical voltage, and more. "Boundary condition" is the specific condition in the element which changes the degrees of freedom in certain parts (or point) of the structure during solution. This means that if any kind of changes in degrees of freedom (such as displacement, rotation, temperature, mean pressure, etc.) are specified in the analysis process, that part is introduced as a boundary condition (whether it is within geometric boundaries or not). Introducing more boundary conditions in the considered geometry leads to the easier and faster solving and analysis of considered FEM model. In general, in most problems, the boundary conditions are required to be determined and defined, whereas some problems (such as static problems) cannot be solved without defining the boundary conditions. "Initial conditions" are defined at a particular time in a problem. These conditions are generally considered at the beginning of the analysis (such as initial velocity and initial temperature).

Moreover, in case of few analysis stages, the conditions at the end of each analysis stage are defined as the initial conditions in the next analysis stage.

The "interaction" involves conditional interactions between different parts of the model geometry. The "contact" and "constrains" are the most common interactions taking place between various parts of the model. The "restrictions," which are defined as the support condition, is considered throughout all analysis steps and these boundary conditions are not revoked under any circumstances. Conditional interactions are of specific types, which occur when certain conditions are met in the model. For example, a contact is made when pressure with the value of more than zero occurs between two parts. In general, when there are more defined interactions in an FEM model, the solving process will be more complicated and time-consuming.

The "connections" is another common type of interaction between different parts of FEM model in which the analytical relationship of one node to another node(s) is determined. This relationship can be the behavior of a linear or nonlinear spring, spherical joint, hinged joint, spot welding, etc., and it can be also used to determine the extent of rupture and the range of changes. Figure 1.2 shows a meshing example of a geometry with regard to interactions and boundary conditions.

### 1.2.6 STEP 6: SOLVING THE FINITE ELEMENT EQUATIONS

Once the boundary conditions are defined properly, the type of analysis that must be specified include of static, dynamic analysis, heat transfer, buckling, natural frequency extraction, etc. The type of elements and the materials should be compatible with the type of analysis. Each type of analysis lead to its own specific output and results. The specific parameters for analysis are needed to define in this step, such as considering small or large displacement or the parameters regarding the nonlinear and plasticity analysis.

Solving of finite element equations, that is, finite element analysis, is conducted after the completion of finite element modeling which leads to determination of nodal displacements resulting due to applied loads to the considered model under the specified boundary conditions.

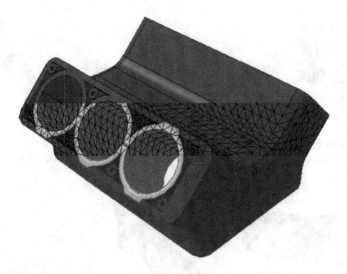

**FIGURE 1.2** Meshing of complex solid geometry with regard to interactions and boundary conditions.

The last step in the analysis of finite element models is reviewing the results and obtaining the outputs. Variables are extracted in different ways which should be defined during modeling process. Some variables are extracted only from nodal results or elements response or both. Furthermore, depending on the type of problem, it is necessary to extract some of the outputs in the current time domain and some through time history results.

### 1.2.7 STEP 7: ADDITIONAL CALCULATIONS

As mentioned in the previous section, the nodal displacement is the only output from finite element analysis. However, many other parameters are required which can be obtained via additional calculations. The strain and stress in the elements are the main concern in design of structures, which can be calculated using nodal displacements. Then it is possible to determine the principal stress and strains to evaluate the capacity of considered members to sustain against applied loads.

The example of implementing various steps of FEM modeling and analysis to simulate an object has been showed in Figure 1.3.

## 1.3 SUMMARY

The above-demonstrated steps form the basis of the main process for finite element modeling and analysis of a solid structure. However, high precision and

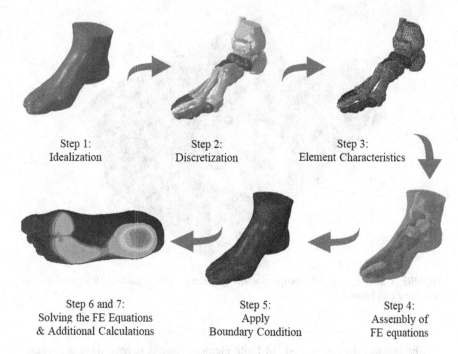

Step 1:                     Step 2:                        Step 3:
Idealization               Discretization            Element Characteristics

Step 6 and 7:                  Step 5:                        Step 4:
Solving the FE Equations        Apply                       Assembly of
& Additional Calculations   Boundary Condition              FE equations

**FIGURE 1.3**  Process of finite element modeling and analysis.

deep consideration are required to define a proper FEM model to obtain accurate results for complex problems with regard to structures and bridges. The examples in the following chapters will provide cognitive insights to the readers to create accurate FEM models for complex structures and to obtain realistic results.

# 2 Abaqus Scripting for Mesh Convergence

## 2.1 INTRODUCTION

The basis for solving finite element programs is the discretization of the component and element size. However, the main challenge in discretization process is to ascertain the appropriate element size to obtain a reliable solution. Therefore, the element(s) size and meshing density is required to be determined reasonably. To establish a proper mesh convergence method, it is required to plot the curve for a critical result (typically stress or displacement) in a specific part or point in the considered model versus mesh density. At least three convergence runs are needed to plot a curve which can be used to indicate that the convergence is achieved or it is desirable to define the optimum convergence constituted by refined mesh. However, if two runs for different mesh density give the same result, the convergence is already achieved, and no convergence curve is necessary. Figure 2.1 shows a typical mesh convergence for displacement.

In this chapter, the process for development of Python scripts to evaluate the mesh convergence has been demonstrated.

## 2.2 PROBLEM DESCRIPTION

The problem consists of a simple rectangular beam that is clamped at one end and subjected to shear loading at the other end (Figure 2.2).

The first step is modeling this example, followed by creation of the scripts.

## 2.3 OBJECTIVES

1. To study writing scripts in Abaqus
2. To familiarize with Abaqus/PDE
3. To investigate mesh independency in Abaqus

## 2.4 MODELING

Run Abaqus/Complete Abaqus environment (CAE) from the start menu and then close the StartSession dialog box (Figure 2.3).

### 2.4.1 PARTS MODULE

The first step is to define a rectangular cubic beam. It should be created as a solid extruded.

DOI: 10.1201/9781003213369-2

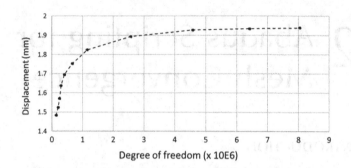

**FIGURE 2.1** Typical mesh convergence.

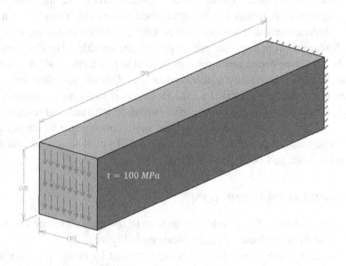

$\tau = 100\ MPa$

**FIGURE 2.2** Problem description.

Double click on Parts in the ModelTree and open the CreatePart dialog box and without changing anything, click Continue to go to the Sketcher (Figure 2.4).

Using Create lines: Rectangle tool, draw a rectangle and using the AddDimension tool determine the dimensions as shown in Figure 2.5.

Click the middle mouse button twice to exit sketcher and open the EditBaseExtrusion dialog box. Consider a depth of 50 and click OK to apply and close the dialog box (Figure 2.6).

### 2.4.1.1 Material Properties

In the next step, the material data should be defined. In the model, the material of the beam has been assumed to be steel in linear elastic.

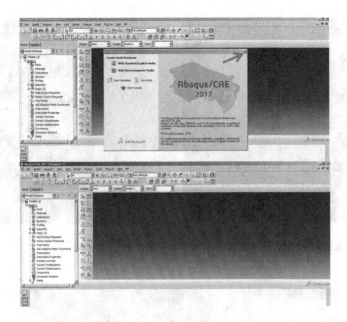

**FIGURE 2.3**   Abaqus/CAE.

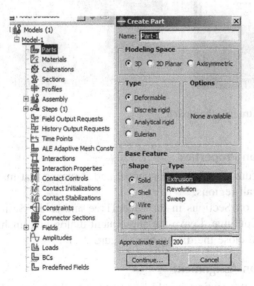

**FIGURE 2.4**   Create Part.

Double click on Materials in the ModelTree and select Mechanical → Elasticity → Elastic and enter 200e3 for Young's modulus and 0.3 for Poisson's ratio, as shown in Figure 2.7.

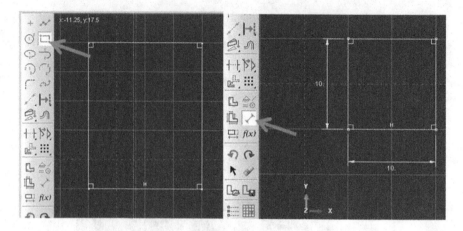

**FIGURE 2.5**  Drawing and dimensioning the sketch of the beam.

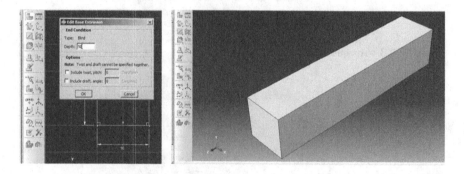

**FIGURE 2.6**  Extruding the sketch.

### 2.4.1.2   Section Properties

The beam is assumed to be homogenous, and it means that material properties are the same in all regions.

Double click on Sections in the ModelTree and select Solid, Homogenous, and then click Continue. Leave the subsequent dialog box unchanged and click OK to apply and close the dialog box (Figure 2.8).

### 2.4.1.3   Section Assignment

In the next step, the section should be assigned to the beam.

Double click on SectionAssignments under Part-1 in the ModelTree. Then click the beam and click the middle mouse button to open the EditSectionAssignment dialog box. Click OK to assign the section to beam and close the dialog box. The beam will be recolored (Figure 2.9).

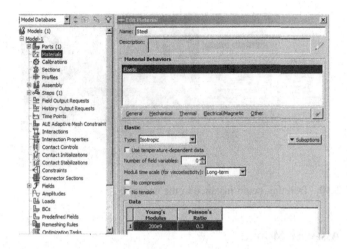

**FIGURE 2.7** Defining material properties.

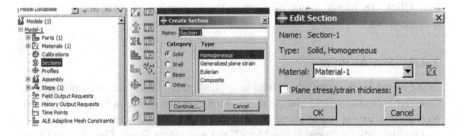

**FIGURE 2.8** Defining section.

**FIGURE 2.9** Assigning section to the beam.

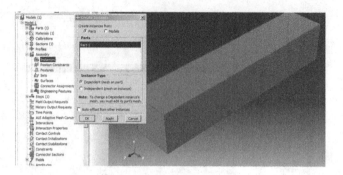

**FIGURE 2.10**   Defining beam instance in the assembly.

### 2.4.2 ASSEMBLY MODULE

At this point, the model should be defined in the computational space in the assembly module.

Double click on Instances under Assembly to open the CreateInstance dialog box. Click OK to add beam instance in the assembly and close the dialog box (Figure 2.10).

### 2.4.3 STEP MODULE

In the example, the static shear load is applied on the beam, while displacements should be considered in the analysis. However, this is a nonlinear problem. Nonlinear analysis is presented in General analysis in Abaqus.

Double click on Steps in the ModelTree and select Static, General, and then click Continue. Toggle on Nlgeom (Nonlinear geometry) and leave all other options as defaults and click OK to apply and close the dialog box (Figure 2.11).

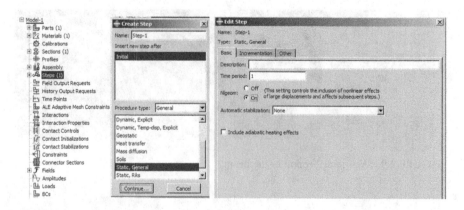

**FIGURE 2.11**   Defining the analysis step.

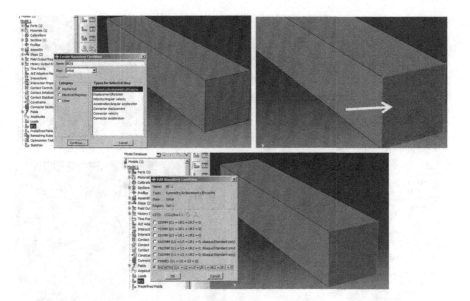

**FIGURE 2.12**   Defining the clamped boundary condition.

## 2.4.4   Boundary Conditions (Load Module)

In this example, only one boundary condition is considered: One end was clamped.

Double click on BCs in the ModelTree to open a corresponding dialog box and select Symmetry/Antisymmetry/Encastre as the type and click Continue. Then rotate the beam so that the end can be selected. Then, click the middle mouse button to open the EditBoundaryCondition dialog box. Constraint all degrees of freedom by toggling Encastre and click OK to apply and close the dialog box (Figure 2.12).

## 2.4.5   Load Module

In this example, a 100 MPa distributed shear was applied, and the direction was set from top to bottom.

Double click on Loads and select SurfaceTraction as a load type and click Continue. Then choose the free end of the beam and click the middle mouse button to open the EditLoad dialog box. Then select edit in the Direction box, and the box will close. Then choose vertex labeled as 1 and then vertex labeled as 2 to define shear load direction. The EditLoad dialog box will open again, enter 100 as the load magnitude and click OK to close the dialog box (Figure 2.13).

## 2.4.6   Mesh Module

The most challenging part of this problem is the mesh size. The main challenge in here is what element size is suitable for the problem. The issue will be

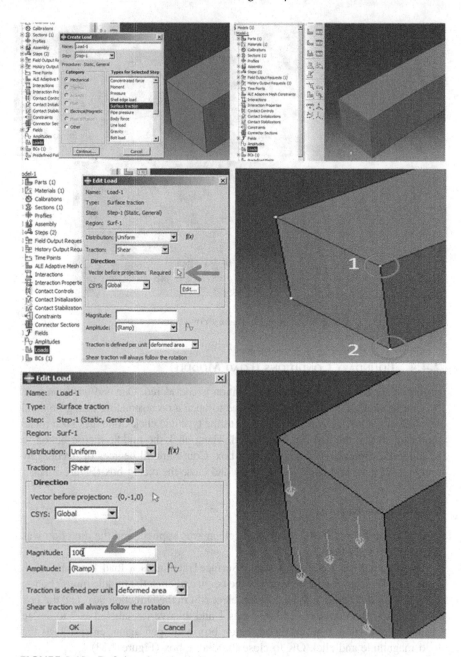

**FIGURE 2.13**  Defining shear load.

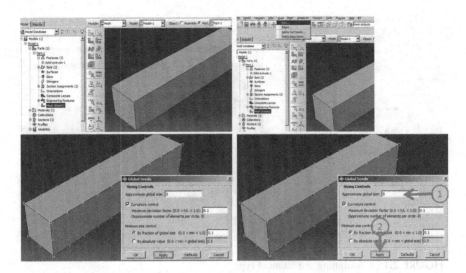

**FIGURE 2.14**   Seeding.

answered using scripting. So, at this point, a simple mesh could be defined; it will be changed with just the script running.

Double click on Mesh (empty) under Part-1 in the ModelTree to open the Mesh module.

To specify the element size, select Seed → Part to open the GlobalSeeds dialog box. Enter 5 as the approximate global size and enter Apply to see the locations of the nodes. Click OK to close the dialog box (Figure 2.14).

Select Mesh → ElementType and choose beam, then click the middle mouse button to open the ElementType box. Change UseDefault to EnhancedStrain in the Hourglass control formulation. This is more suitable for calculating bending due to eliminating twist instabilities caused by one integration point. To assign the element type, click OK, and close the dialog box (Figure 2.15).

Select Mesh → Part and click Yes at the bottom of the page to generate the mesh (Figure 2.16).

## 2.5   ANALYSIS: JOB MODULE

In the next step, a job should be defined to solve the problem in the solver. Before that, the assembly should be regenerated again.

Right click on Instances under Assembly and select Regenerate (Figure 2.17).

Double click on Jobs in the ModelTree and then name the job, click Continue to open the EditJob dialog box. Then click OK to create the job and close the dialog box (Figure 2.18).

Finally, save the model as Beam.cae and close Abaqus/CAE.

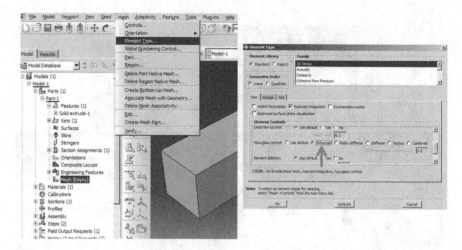

**FIGURE 2.15** Assigning the element type.

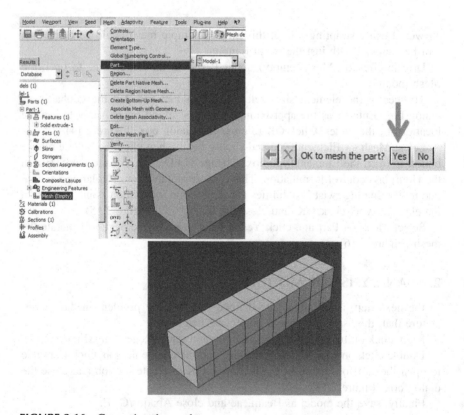

**FIGURE 2.16** Generating the mesh.

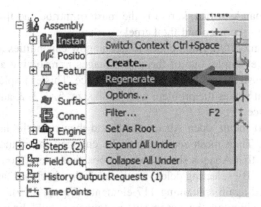

**FIGURE 2.17**   Assembly regeneration.

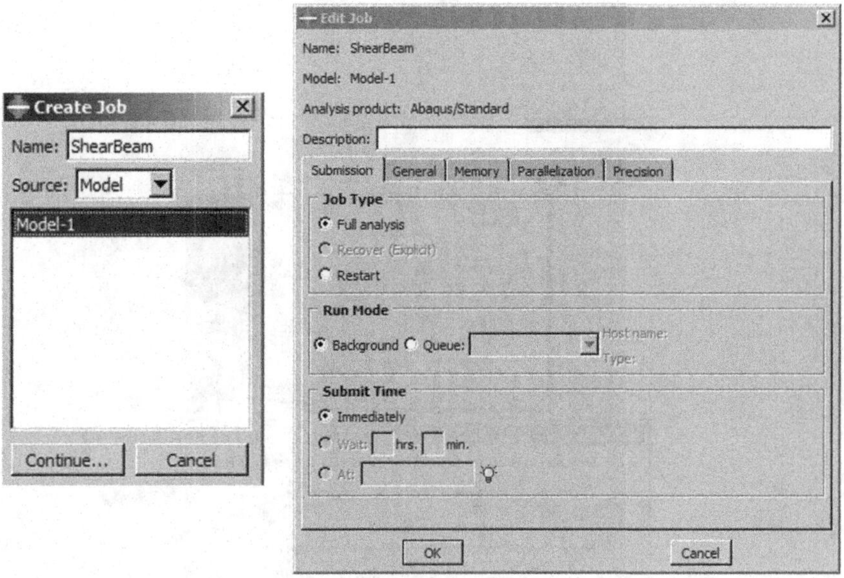

**FIGURE 2.18**   Creating a job.

Using the Abaqus/CAE user interface to create a geometric model and view results, the all executive comments are created and stored in the form of codes following each operation. These commands contain information on how to create a geometric model and change the settings in all windows that are used in each module. The graphic user interface (GUI) or Abaqus interface will create all the commands in a object oriented programming language, called Python. All commands created by the central kernel or the Abaqus kernel are translated and based on the settings created. The user is able to view the visual representation of the issued command. In fact, kernel is the mastermind behind the Abaqus/CAE and GUI between the user and the kernel.

The user can modify Python codes to define new capabilities and advantages by Abaqus. For example, one of these advantages is resolving a problem with varying mesh density to evaluate mesh convergence. This example shows how to write a script that solves the problem with mesh size: 5, 4, 3, 2, and 1 to evaluate mesh convergence.

From the start menu, open Abaqus/CAE and then open the last model database: Beam.cae. Then mesh seeds should be changed to another size, for example, 4, before that Abaqus shows a dialog box that asks if you wish to delete the current mesh. By selecting delete mesh, the dialog is closed, and a mesh part should be created again containing 117 elements (Figure 2.19).

As meshing changed the assembly, instances should be regenerated by right-clicking on them. After that, a Job should be created and submitted (Figure 2.20).

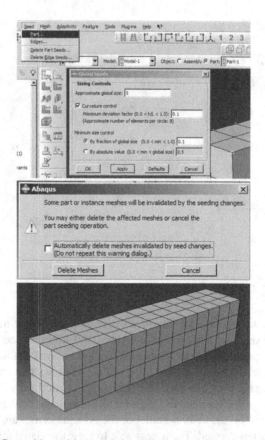

**FIGURE 2.19** Remeshing with a new element size.

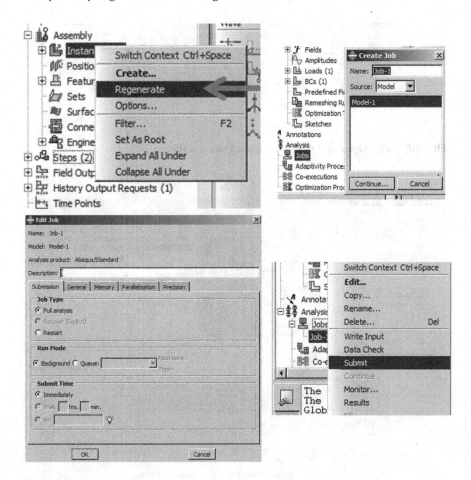

**FIGURE 2.20** Submitting the job including the new mesh.

After the analysis is completed, close Abaqus/CAE without saving anything.

Since Abaqus/CAE runs until it is closed, all items are recorded as Python scripts by Abaqus/PDE and can be edited for another analysis. All changes are stored in the Abaqus.rpy file, which is generally located in c:/temp directory.

Open Abaqus.rpy in the notepad.exe editor, as shown in Figure 2.21.

Save the file as mesh.py and close it. The *.py extension is used for Python scripts.

Run Abaqus/CAE and select File. Abaqus manages and modifies all Python commands. The recent Python file—mesh.py—should be opened in Abaqus/PDE, as shown in Figure 2.22.

Modify the script as described below:

**FIGURE 2.21**   Opening *.rpy file contains Python scripts.

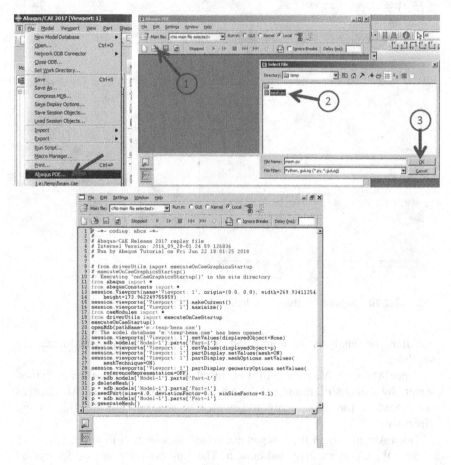

**FIGURE 2.22**   Opening the mesh.py file, a script that is included to modify mesh and submit a job.

Remove all commands that started with "#" which mention all analysis details such as time, date[1] (Figure 2.23).

Remove all "session.viewports" commands, which mention all orientations of the viewport and zoom in/zoom out done by the user which does not affect the script (Figure 2.24).

Remove all commands that contain "driverUtils" and "caeModules" that mention Abaqus/CAE as they are not necessary for the script (Figure 2.25).

Remove all duplicated statements that start with "p.", "a.", "p=", or "a=" as the final script performs every task just once (Figure 2.26).

Define the new statement as "MeshSizes" after "p=mdb.models..." statement. By doing so, an integer can be defined, and the meshing operator in Abaqus/CAE will refer to it in every execution [run]. The statement should be written using a unique interface, as shown below and as illustrated in Figure 2.24

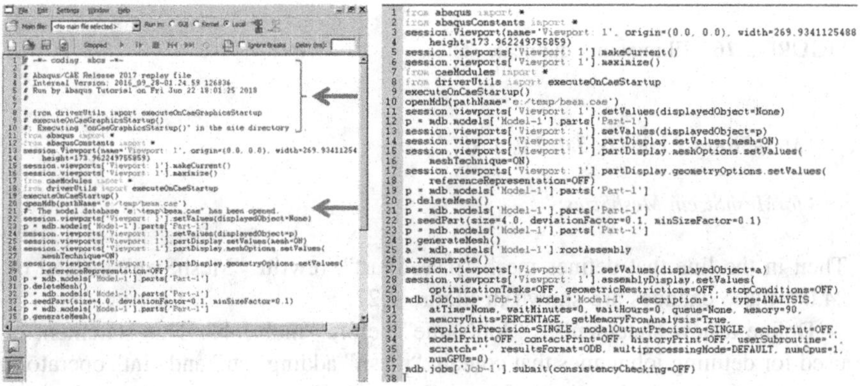

**FIGURE 2.23** Eliminating anything that starts with "#".

**FIGURE 2.24** Eliminating states that start with "session.viewports".

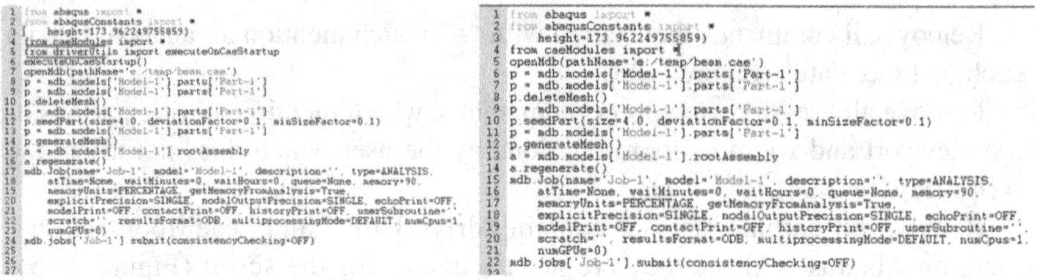

**FIGURE 2.25** Eliminating commands that contain "driverUtils" and "caeModules".

**FIGURE 2.26** Eliminating duplicate states.

"MeshSizes= [5.0, 4.0, 3.0, 2.0, 1.0]"

"forMeshSizein MeshSizes:"

Then in the line that defines mesh "p.seedPart", rewrite "MeshSize" instead of "4.0"; these changes are shown in Figure 2.27.

Define a new statement as "JobName" before "mdb.Job". This statement is used for defining job names that contain "mesh" adding "str" and "int" operators for extracting a number that refers to the mesh size to generate new jobs. All

```
1  from abaqus import *
2  from abaqusConstants import *
3    height=173.962249755859)
4  from caeModules import *
5  openMdb(pathName='e:/temp/beam.cae')
6  p = mdb.models['Model-1'].parts['Part-1']
7  p.deleteMesh()
8  p.seedPart(size=4.0, deviationFactor=0.1, minSizeFactor=0.1)  ]
9  p.generateMesh()
10 a = mdb.models['Model-1'].rootAssembly
11 a.regenerate()
12 mdb.Job(name='Job-1', model='Model-1', description='', type=ANALYSIS,
13     atTime=None, waitMinutes=0, waitHours=0, queue=None, memory=90,
14     memoryUnits=PERCENTAGE, getMemoryFromAnalysis=True,
15     explicitPrecision=SINGLE, nodalOutputPrecision=SINGLE, echoPrint=OFF,
16     modelPrint=OFF, contactPrint=OFF, historyPrint=OFF, userSubroutine='',
17     scratch='', resultsFormat=ODB, multiprocessingMode=DEFAULT, numCpus=1,
18     numGPUs=0)
19 mdb.jobs['Job-1'].submit(consistencyChecking=OFF)

MeshSizes= [5.0, 4.0, 3.0, 2.0, 1.0]
for MeshSize in MeshSizes:
    p.deleteMesh()
    p.seedPart(size=MeshSize, deviationFactor=0.1, minSizeFactor=0.1)
```

**FIGURE 2.27** Adding the "Meshsize" statement in the script.

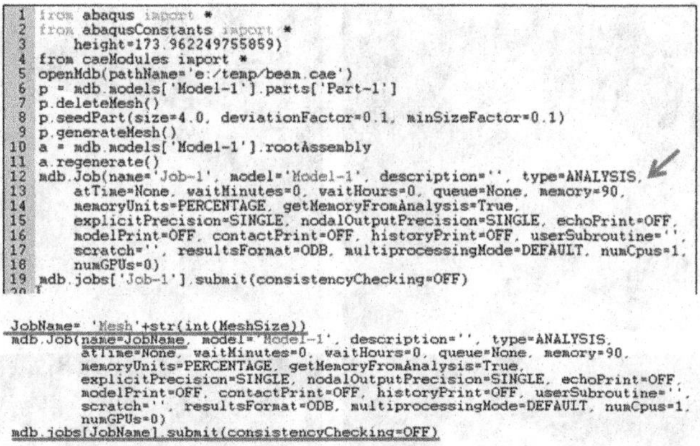

from abaqus import *
from abaqusConstants import *
    height=173.962249755859)
from caeModules import *
openMdb(pathName='e:/temp/beam.cae')
p = mdb.models['Model-1'].parts['Part-1']
p.deleteMesh()
p.seedPart(size=4.0, deviationFactor=0.1, minSizeFactor=0.1)
p.generateMesh()
a = mdb.models['Model-1'].rootAssembly
a.regenerate()
mdb.Job(name='Job-1', model='Model-1', description='', type=ANALYSIS,
    atTime=None, waitMinutes=0, waitHours=0, queue=None, memory=90,
    memoryUnits=PERCENTAGE, getMemoryFromAnalysis=True,
    explicitPrecision=SINGLE, nodalOutputPrecision=SINGLE, echoPrint=OFF,
    modelPrint=OFF, contactPrint=OFF, historyPrint=OFF, userSubroutine='',
    scratch='', resultsFormat=ODB, multiprocessingMode=DEFAULT, numCpus=1,
    numGPUs=0)
mdb.jobs['Job-1'].submit(consistencyChecking=OFF)
JobName= 'Mesh'+str(int(MeshSize))
mdb.Job(name=JobName, model='Model-1', description='', type=ANALYSIS,
    atTime=None, waitMinutes=0, waitHours=0, queue=None, memory=90,
    memoryUnits=PERCENTAGE, getMemoryFromAnalysis=True,
    explicitPrecision=SINGLE, nodalOutputPrecision=SINGLE, echoPrint=OFF,
    modelPrint=OFF, contactPrint=OFF, historyPrint=OFF, userSubroutine='',
    scratch='', resultsFormat=ODB, multiprocessingMode=DEFAULT, numCpus=1,
    numGPUs=0)
mdb.jobs[JobName].submit(consistencyChecking=OFF)

**FIGURE 2.28** Adding the "JobName" statement.

File Edit Settings Window Help
Main file: e:/temp/mesh.py
Run in: GUI Kernel Local
Stopped
Ignore Breaks
Delay (ms):
Play
from abaqus import *
from abaqusConstants import *
from caeModules import *
openMdb(pathName='e:/temp/beam.cae')
p = mdb.models['Model-1'].parts['Part-1']
MeshSizes= [5.0, 4.0, 3.0, 2.0, 1.0]
for MeshSize in MeshSizes:
    p.deleteMesh()
    p.seedPart(size=MeshSize, deviationFactor=0.1, minSizeFactor=0.1)
    p.generateMesh()
    a = mdb.models['Model-1'].rootAssembly
    a.regenerate()
    JobName= 'Mesh'+str(int(MeshSize))
    mdb.Job(name=JobName, model='Model-1', description='', type=ANALYSIS,
        atTime=None, waitMinutes=0, waitHours=0, queue=None, memory=90,
        memoryUnits=PERCENTAGE, getMemoryFromAnalysis=True,
        explicitPrecision=SINGLE, nodalOutputPrecision=SINGLE, echoPrint
        modelPrint=OFF, contactPrint=OFF, historyPrint=OFF, userSubrouti
        scratch='', resultsFormat=ODB, multiprocessingMode=DEFAULT, numC
        numGPUs=0)
    mdb.jobs[JobName].submit(consistencyChecking=OFF)
    mdb.jobs[JobName].waitForCompletion()

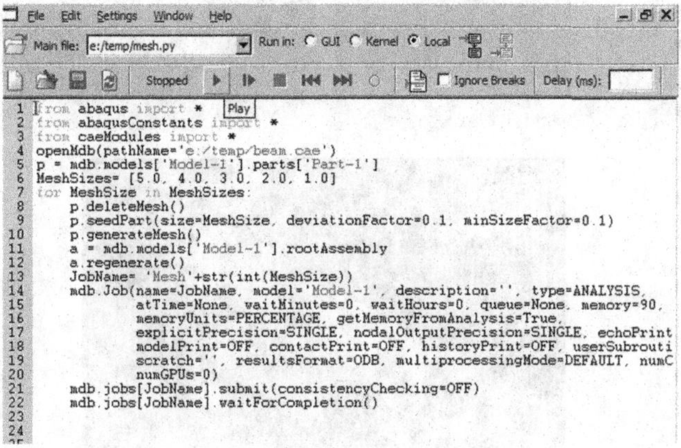

**FIGURE 2.29** Adding the "waiting" command.

File Edit Settings Window Help
Main file: e:/temp/mesh.py
Run in: GUI Kernel Local
Stopped
Ignore Breaks
Delay (ms):
Play
from abaqus import *
from abaqusConstants import *
from caeModules import *
openMdb(pathName='e:/temp/beam.cae')
p = mdb.models['Model-1'].parts['Part-1']
MeshSizes= [5.0, 4.0, 3.0, 2.0, 1.0]
for MeshSize in MeshSizes:
    p.deleteMesh()
    p.seedPart(size=MeshSize, deviationFactor=0.1, minSizeFactor=0.1)
    p.generateMesh()
    a = mdb.models['Model-1'].rootAssembly
    a.regenerate()
    JobName= 'Mesh'+str(int(MeshSize))
    mdb.Job(name=JobName, model='Model-1', description='', type=ANALYSIS,
        atTime=None, waitMinutes=0, waitHours=0, queue=None, memory=90,
        memoryUnits=PERCENTAGE, getMemoryFromAnalysis=True,
        explicitPrecision=SINGLE, nodalOutputPrecision=SINGLE, echoPrint
        modelPrint=OFF, contactPrint=OFF, historyPrint=OFF, userSubrouti
        scratch='', resultsFormat=ODB, multiprocessingMode=DEFAULT, numC
        numGPUs=0)
    mdb.jobs[JobName].submit(consistencyChecking=OFF)
    mdb.jobs[JobName].waitForCompletion()

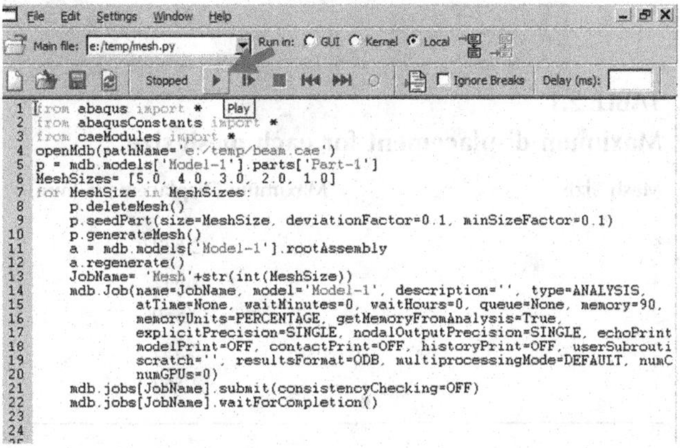

**FIGURE 2.30** Play script.

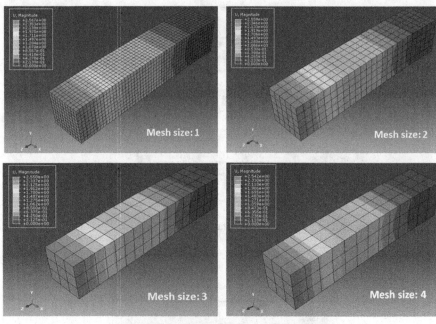

**FIGURE 2.31**  Result for different mesh sizes.

**TABLE 2.1**
**Maximum displacement for each mesh size**

| Mesh Size | Maximum Displacement (mm) |
|-----------|---------------------------|
| 5         | 2.532                     |
| 4         | 2.542                     |
| 3         | 2.550                     |
| 2         | 2.559                     |
| 1         | 2.567                     |

following statements that refer to "Job-1" should be changed according to the statement ("JobName"). All changes required for the statement interface are shown in Figure 2.28.

Define the waiting command: This command keeps the analysis running and prevents new analysis from starting before the completion of the current one. Its interface has been mentioned below. Note that the command should be the last line in Abaqus/PDE (Figure 2.29).

*"mdb.jobs[JobName].waitForCompletion()"*

Finally, the script should be saved in the mesh.py file. Then, run it using the Play button at the top of the Abaqus/PDE (Figure 2.30). When the script is running, various activities take place in Abaqus/CAE, including defining models with different mesh size and solving the problem according to the size. After finishing the analysis, a new mesh density will be generated and analyzed.

After all the jobs have been completed, the result is displayed. For example, a comparison of maximum displacement in five models, as shown in Figure 2.31.

The results summary and mesh evaluation are shown in Table 2.1. The values show the refining mesh size, where the difference between the results is less. It also shows the minimum mesh size that is likely to be qualified for mesh independency.

## NOTE

1 Note that upper and lowercase and spaces are very important in scripting.

# 3 Analysis of Steel Rectangular Plate with Circular Hole

## 3.1 INTRODUCTION

In this example, a rectangular plate with a circular hole is subjected to horizontal force. Then, the effect of stretching of the plate on the stress distributions around the hole is analyzed using the finite element method.

## 3.2 PROBLEM DESCRIPTION

A rectangular plate with a circular hole at its center is subjected to uniform horizontal stress of 10 kN/m$^2$ along both sides. It should be noticed that the plate has no boundary condition, and it is a free movin g body as it is not fixed in position. However, in finite element analysis, it should have introduced at least one boundary condition, to process the analysis. When it is possible to create the symmetry condition in the model, it should be considered in the model to reduce the computational time in the analysis. Therefore, the plate is modeled as a quarter where its symmetry condition will play the role of boundary conditions in this example. The cut edge will be designed fully constrained, having a uniform distributed load on the other end, as shown in Figure 3.1. In this case, due to the symmetry in geometry concerning the equal applied loads, a quarter of the model can be modeled. Therefore, analysis of plate in a state of plane stress has been carried out, and the stress sdistribution and deformation of a plate are investigated.

Modulus of elasticity = 20,000 kN/cm$^2$
Poisson's ratio = 0.3

## 3.3 OBJECTIVES

1. To determine the effect of load applied to the plate on the deformation behavior of the hole
2. To investigate the stress distribution of a rectangular plate with a circular hole

DOI: 10.1201/9781003213369-3

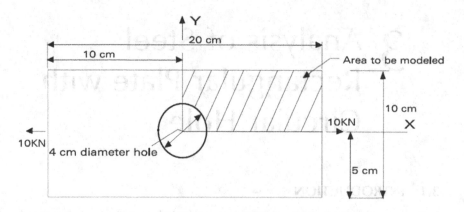

Analyze one–quarter of the above steel plate.
Thickness = 0.5 cm
Material: Steel

**FIGURE 3.1**   Plate with hole.

## 3.4  MODELING

### 3.4.1  PART MODULE

This module allows for creating the geometry required for the problem. To create
a 3D geometry, a 2D section should be created first and then manipulated to
obtain the solid geometry.

#### 3.4.1.1   Create a New Model Database

Start Abaqus/CAE from programs in the Start menu.

Select Create Model Database from the Start Session dialog box that appears
(see Figure 3.2).

When the Part module has finished loading, it displays the Part module
toolbox in the left side of the Abaqus/CAE main window. Each module displays
its own set of tools in the module toolbox (Figure 3.2).

#### 3.4.1.2   Create a New Model Database and a New Part

From the main menu bar, select Part→Create to create a new part.

The Create Part dialog box appears. Use the Create Part dialog box to name
the part and to select its modeling space, type, and base feature and to set the
approximate size. The name of the part may be edited once it has been created,
but the modeling space, type, or base feature cannot be changed.

Name the part *plateWithHole*. Choose a 2D planar, deformable body type, and
Shell as the base feature (see Figure 3.3).

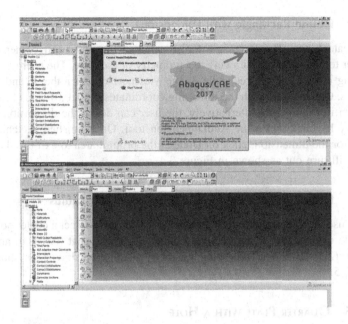

**FIGURE 3.2**   Getting started.

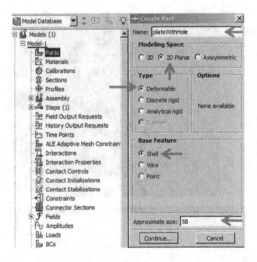

**FIGURE 3.3**   Create a new model database and create a new part.

Enter an approximate size of 50. The value entered in the approximate size text field at the bottom of the dialog box sets the approximate size of the new part.

Click [Continue] to exit the CreatePart dialog box.

### 3.4.1.3 Define Rectangle with Dimensions

Use Create Lines: The Rectangle tool located in the upper left corner of the Sketcher toolbox to begin drawing the geometry of the plate. The user can select a starting corner for the rectangle at the viewport or enter the X and Y coordinates. Create a line with the following coordinates: (0.0, 0.0), and (10.0, 5.0) as shown in Figure 3.4. Alternatively, the user can define the dimension of the geometry by clicking on the AddDimension tool. Once finished sketching the section for the dimension, click on Done at the prompt area to exit the sketcher, which will turn out as shown in Figure 3.4.

### 3.4.1.4 Draw Circle to Define Cut and Dimension Radius

Finally, use the CreateCircle: Center and Perimeter tool. Select a center point for the circle or enter the X and Y coordinates with a center at (0.0, 0.0) and one point at (2.5, 0.0). The following geometry should be drawn, as shown in Figure 3.5.

### 3.4.1.5 QUARTER PLATE WITH A HOLE

Click Done at the prompt area to exit the Sketcher (see Figure 3.6).

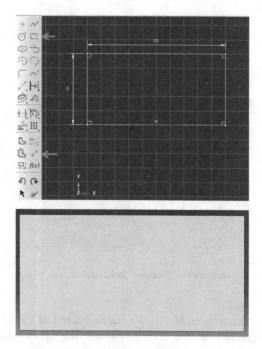

**FIGURE 3.4** Define the rectangle and dimensions.

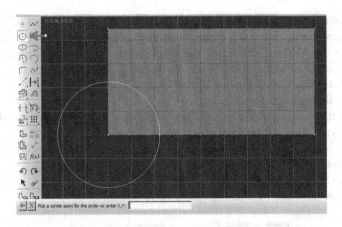

**FIGURE 3.5**   Extrude cut to create a hole.

**FIGURE 3.6**   Draw circle to define cut and dimension radius.

Save the model in a model database file: From the main menu bar, select File→Save. The Save Model Database appears as a dialog box.

Type a name for the new model database in the File Name field and click OK.

### 3.4.2   PROPERTY MODULE

In this module, the material properties for the analysis should be defined by assigning those properties to the available parts.

Note: If the Done button in the prompt area does not appear, right click on the viewport until it appears.

#### 3.4.2.1   Material Properties

The Property module is used to create a material and to define its properties. In this problem, all the members of the frame are made of steel and are assumed to be linear elastic with Young's modulus of 20,000 kN/cm$^2$ and Poisson's ratio of 0.3. Thus, a single linear elastic material is created with these properties. To define a material:

Under the Model Tree, select Property, to open the Property module. The cursor changes to an hourglass while the Property module is loaded.

The Edit Material dialog box appears.

Name the material *Steel*.

From the material editor's menu bar, click Mechanical→Elasticity→Elastic. The software displays the Elastic data form (see Figure 3.7).

Enter the value of 20,000 for Young's modulus and 0.3 for Poisson's ratio in the respective cells (see Figure 3.8).

Click OK to exit the material editor.

Keep the changes by clicking on the Save button.

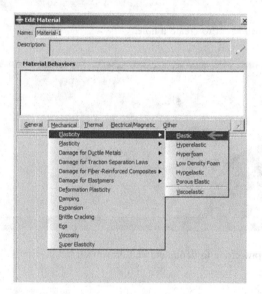

**FIGURE 3.7**  Material definition.

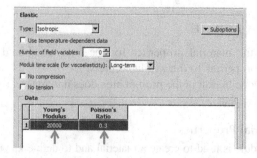

**FIGURE 3.8**  Material properties.

### 3.4.2.2   Section Properties

The section properties of a model are defined by creating sections in the Property module. Once the section has been created, use one of the following two methods to assign the section to the part:

Select the region from the part and assign the section to the selected region, or use the Set toolset to create a homogenous set containing the region and assign the section to the set.

To define a plate section as shown in Figure 3.9 the following process should be fulfilled.

From the main menu bar, select Section→Create and the Create Section dialog box appears.

In the Create Section dialog box:

Name the section *plate Section*.

In the Category list, select Solid.

In the Type list, select Homogenous.

Click Continue. The Edit Section dialog box appears.

In the Edit Section dialog box:

Accept the default selection of Steel for the Material associated with the section. If other materials have previously been defined, click the arrow next to the Material text box and scroll through the Material to view a list of available materials and assign it to the section.

In the Plane stress/strain thickness field, enter the value as 0.5.

Click OK.

### 3.4.2.3   Assign Plate Section to the Part

Next, assign the defined section to the corresponding part. The Assign menu is used in the Property module to assign the section *plateSection* to the plate. Assign the section to the plate, as shown in Figure 3.10.

From the main menu bar, select Assign→Section. Abaqus/CAE displays prompts in the prompt area to guide the user through the procedure.

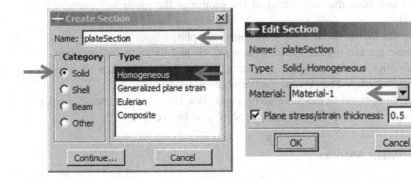

**FIGURE 3.9**   Section properties.

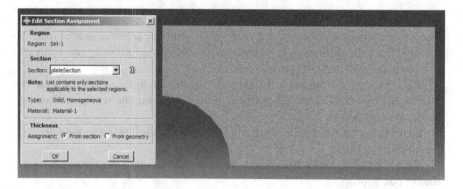

**FIGURE 3.10**   Plate section.

**FIGURE 3.11**   Plate section assignment.

Alternatively, expand the menu under the *plateWithHole* and double click on the Section Assignments.

Select the entire part as the region which the section will be applied.

Click and hold the left button of the mouse at the upper left corner of the viewport.

Drag the mouse to create a box around the plate.

Release the left mouse button. Abaqus/CAE highlights the entire plate.

Right click on the viewport or click Done in the prompt area to accept the selected geometry. The Assign Section dialog box appears.

In Section, scroll to the plate section and click OK. The part changes color to green once the section is assigned, as shown in Figure 3.11.

### 3.4.3 Mesh Module

A finite element mesh is generated in this module. Abaqus/CAE uses a number of different meshing techniques. The default meshing technique assigned to the model is indicated by the color of the model which is displayed when the Mesh module is opened. If Abaqus/CAE displays the model in orange, it cannot be meshed without the assistance of the user. This command is used to mesh the whole structure to small and equal parts and elements.

### 3.4.3.1  Mesh: Seed the Part (1 cm Elements)

The Mesh module is used to generate the finite element mesh. The meshing technique used to create the mesh, the element shape, and the element type, should be chosen. In this section, a particular Abaqus element type is assigned to the model. The element type assignment can also be postponed until the mesh is created. The Plane stress elements will be used to model the plate. To assign an Abaqus element type:

In the Part Module list, expand the menu under the *plateWithHole* and double click on the Mesh to open the Mesh module.

At the context bar, click Part, to unclick the assembly.

From the main menu bar, select Mesh→Element Type.

In the viewport, select the entire frame as the region to be assigned with an element type. In the prompt area, click Done. The Element Type dialog box appears, as shown in Figure 3.12.

In the dialog box, select the following:

- Standard as the Element Library selection (the default)
- Linear as the Geometric Order (the default)
- Plane stress as the Family of elements
- Unclick the reduced integration box (**NEVER** use *reduced integration*)

In the lower portion of the dialog box, examine the element shape options. A brief description of the default element selection is available at the bottom of each tabbed page.

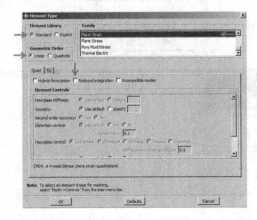

FIGURE 3.12  Selecting the element type.

Click OK to assign the element type and to close the dialog box (see Figure 3.12).

Then the mesh can be created. Meshing is basically a two-stage operation that includes first seeding the edges of the part instance followed by mesh in the part instance. Select the number of seeds based on the desired element size or on the number of elements that are required along an edge, and Abaqus/CAE places the nodes of the mesh at the seeds whenever possible. For manual meshing, click on Seed Edges. Then, select the edges that the user wants to mesh. Next, click Mesh Part Instance.

### 3.4.3.2 Seed and Mesh the Model

From the main menu bar, select Seed→Part to seed the part instance.

Alternatively, select the Seed Part on the upper left corner of the meshing toolbox. The Global Seeds dialog box will appear.

Type the appropriate value for the approximate global size of the mesh elements. Specify an element size of 1.0 for this example.

Click OK to accept the seeding (see Figure 3.13).

Note: More control of the resulting mesh can be gained by seeding each edge of the part instance individually. However, it is not necessary for this example. The prompt area displays the default element size that Abaqus/CAE will use to seed the part instance. This default element size is based on the size of the part instance. A relatively large seed value will be used so that only one element will be created per region.

Abaqus/CAE offers a variety of meshing techniques to mesh models with different topologies. Various meshing techniques provide different levels of automation and user control. There are three types of mesh-generation techniques available in Abaqus/CAE as follows:

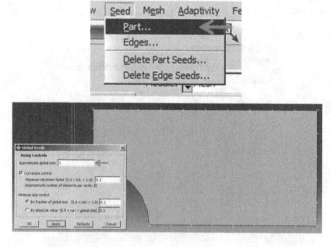

**FIGURE 3.13**   Assign the approximate global size for the mesh elements.

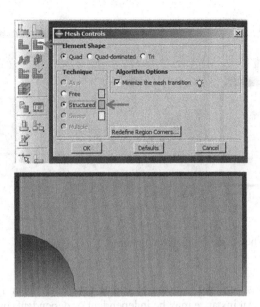

**FIGURE 3.14** Choose element shape and the meshing technique option.

Structured meshing applies pre-established mesh patterns to particular model topologies. To use this technique, complex models must generally be partitioned into simpler regions.

Swept meshing extrudes an internally generated mesh along a sweep path or revolves it around an axis of revolution. Like structured meshing, swept meshing is limited to models with specific topologies and geometries.

Free meshing is the most flexible meshing technique as it uses no pre-established mesh patterns and can be applied to almost all model shapes.

Select the AssignMeshControls tool at the meshing toolbox, and the MeshControls dialog box appears.

Choose Quad as the element shape. Then, ChooseStructured meshing technique, as shown in Figure 3.14.

From the main menu bar, select Mesh→Instance to mesh the part instance or select the MeshPartInstance at the upper left corner of the meshing toolbox.

Select the part instances to be meshed. Once finished selecting, click Yes in the prompt area to confirm the mesh of the part instance. Once meshed, the plate changes color to blue, and the meshed geometry is shown in Figure 3.15.

### 3.4.4 Assembly Module

Each part is oriented in its own coordinate system and is independent of the other parts in the model. Although a model may contain many parts, it contains only one assembly. Define the geometry of the assembly by creating instances of a

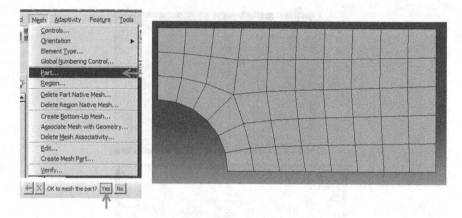

**FIGURE 3.15** Choose mesh algorithm options.

part and then positioning the instances relative to each other in a global co-ordinate system. An instance may be independent or dependent. Independent part instances are meshed individually, while the mesh of a dependent part instance is associated with the mesh of the original part.

### 3.4.4.1 Assemble Part Instances into the Model

In the Module list located under the toolbar, click Assembly to open the Assembly module.

From the main menu bar, select Instance→Create or select the Create Instance at the upper left of the assembling toolbox. The Create Instance dialog box appears, as shown in Figure 3.16.

Note: In Abaqus, the user can create many parts and assemble them to form a model. The user can also create many instances from one part. For example, in a composite structure, the user does not need to draw all the bolted shear

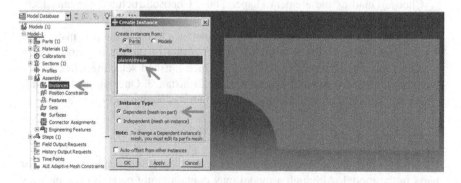

**FIGURE 3.16** Assemble part instances into a model.

connectors. If they are similar, drawing one is sufficient. The others can be created using the linear pattern from the first one by defining the number of instances and the offset from one to another.

In the dialog box, choose Dependent (mesh on the part) as the instance type.

In the considered case, we only have one part: *plateWithHole*, select it, and click OK.

### 3.4.5 STEP MODULE

After the assembly is completed, then the configuration of the analysis should be defined. In this simulation analysis, we are interested in identifying the static response of the plate to a 10 kN/m$^2$ load applied at the adjacent sides.

Abaqus/CAE generates the initial step automatically. However, the rest of the analysis step should be defined by the user as well as the requested output for any steps in the analysis. There are two kinds of analysis steps in Abaqus: general analysis steps, which can be used to analyze the linear or nonlinear response, and linear perturbation steps, which can be used only to analyze linear problems. However, only general analysis steps are available in Abaqus/Explicit. Next, create a static, general that follows the initial step of the analysis.

#### 3.4.5.1   Create an Analysis Step

In the next step, that is, after the assembly of the model has been created, move to the Step module to configure the required analysis. In this simulation analysis, the aim is to evaluate the static response of the plate to a 10 kN/m$^2$ pressure applied at the end of the plate, with the left end fully constrained. This is a single event, so only a single analysis step is needed for the simulation. Thus, the analysis will consist of two steps:

- The initial step in which boundary conditions that constrain the end of the plate are applied
- Analysis step, in which a distributed load at the other end of the plate is applied

Abaqus/CAE generates the initial step automatically, but the Step module needs to be operated by the user to create the analysis step. The Step module also allows the user to request output (field output & history output) for each step in the analysis.

In the Module list located under the toolbar, click Step to open the Step module.

From the main menu bar, select Step→Create to create a step. The Create Step dialog box appears with a list of all general procedures and a default step named *Step-1* (see Figure 3.17).

Select General as the Procedure type.

Scroll through the available list, select Static, General, and click on Continue.

Next, the EditStep dialog box appears.

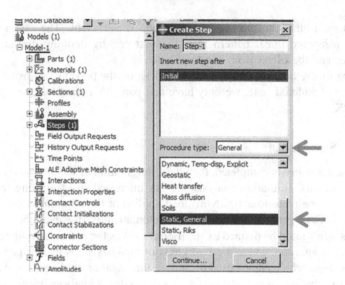

**FIGURE 3.17**   Analysis step.

The Basic tab is selected by default. In the Description field, type *This is a load step where we apply the 10 kN force at the end* (see Figure 3.18).

Click the Incrementation tab and the Other tab to see its contents and accept the default values provided for the step.

Click OK to create the step and to exit the Edit Step dialog box.

### 3.4.6 Load Module

The prescribed conditions, such as loads and boundary conditions, are step dependent, which means that the user needs to specify the steps in which they become active. As the steps in the analysis have been defined, the Load module can be used to define the prescribed conditions. In this model, the left edge of the plate is fully constrained and cannot move in any direction. However, due to the symmetry of the x and y axes, one-fourth of the plate is modeled, with the boundary conditions shown in Figure 3.19.

To apply boundary conditions to the plate, follow the steps as shown below:

In the Module list located under the toolbar, click Load to open the Load module.

From the main menu bar, select BC→Create. The Create Boundary Condition dialog box appears.

In the Create Boundary Condition dialog box:

Name the boundary condition *leftBC*.

From the list of steps, select Initial as the step in which the boundary condition will be activated. All the mechanical boundary conditions specified in the

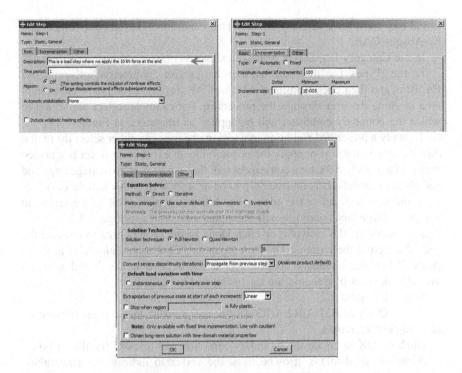

**FIGURE 3.18**   Time options, Nlgeom, and incrementation.

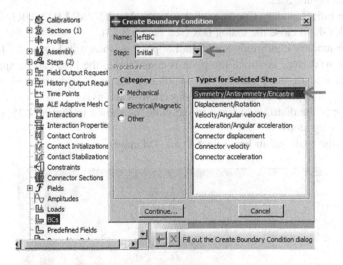

**FIGURE 3.19**   Left symmetric boundary condition.

Initial step must have zero magnitude. This condition is enforced automatically by Abaqus/CAE.

In the Category list, accept Mechanical as the default category selection.

In the Types for the SelectedStep list, select Symmetry/Antisymmetry/ Encastre, and then click Continue. Abaqus/CAE displays prompts in the prompt area to guide the user through the procedure. For example, select the region to which the boundary condition will be applied as illustrated in Figure 3.19.

To apply a prescribed condition to a region, the user can either select the region directly in the viewport or apply the condition to an existing set (a set is a named region of a model). Sets are a convenient tool that can be used to manage large and complicated models. It is unnecessary to have more sets in this simple model.

In the viewport, select the edge at the left of the plate. This is the region to which the boundary condition will be applied, as shown in Figure 3.20.

Right-click on the viewport or click Done in the prompt area to indicate the end of selecting the regions. The Edit Boundary Condition dialog box then appears. When the boundary condition is being defined in the initial step, all available degrees of freedom are unconstrained by default.

In the dialog box:

Select XSYMM (U1=UR2=UR3=0) since all translational degrees of freedom need to be constrained.

Click on OK to create the boundary condition and to close the dialog box.

Abaqus/CAE displays arrowheads at the vertex to indicate the constrained degrees of freedom (see Figure 3.21).

Repeat the steps for the lower edge of the plate with the name *lowerBC,* as shown in Figure 3.22.

In the Edit Boundary Condition dialog box, select YSYMM (U2=UR1=UR3=0) and click OK to close the dialog box (see Figure 3.23).

The boundary conditions applied all appear, as shown in Figure 3.24.

In the next step, in which the plate is constrained, the load can be applied to the other end of the plate. In this simulation analysis, a distributed force of 10 kN/m$^2$ is applied in the negative direction of the axis.

To apply the distributed force to the plate, as shown in Figure 3.25, perform the following steps:

From the main menu bar, select Load→Manager.

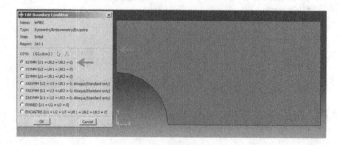

FIGURE 3.20   Edit the boundary condition dialog box for fixed support.

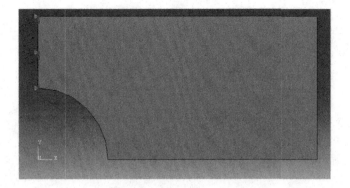

**FIGURE 3.21**  Select edge and BC options.

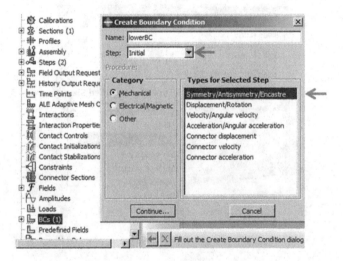

**FIGURE 3.22**  Lower symmetric BC.

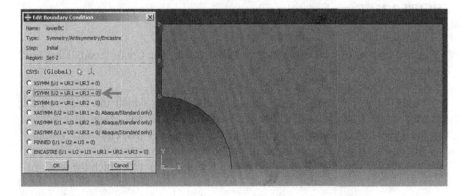

**FIGURE 3.23**  Select edge and BC options.

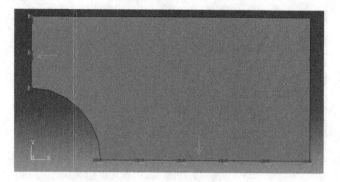

**FIGURE 3.24** All boundary conditions shown (applied in initial step).

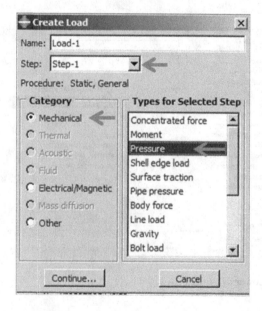

**FIGURE 3.25** Create load dialogue box.

At the bottom of the Load Manager, click Create. The Create Load dialog box appears.

In the CreateLoad dialog box:

From the list of steps, select Step-1 as the step in which the load will be exerted. In the Category list, select Mechanical as the default category selection. In the Types for the SelectedStep list, select Pressure. Then, click on Continue.

Abaqus/CAE displays prompts in the prompt area to guide the user through the procedure. The user is asked to select a region or point to which the load will

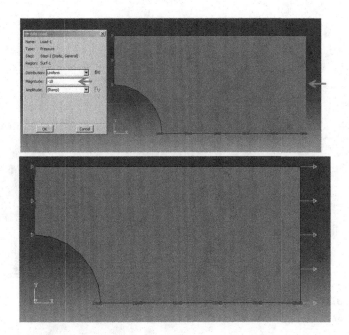

**FIGURE 3.26** Defining load.

be applied. As with the boundary conditions, the region to which the load will be applied can be selected either directly in the viewport or from a list of existing sets. Select the region directly in the viewport.

In the viewport, select the right edge of the plate as the region where the load will be applied.

Click on the viewport or click Done in the prompt area to finish selecting the regions. The EditLoad dialog box appears.

In the dialog box (see Figure 3.26):

Click OK to create the load and to close the dialog box.

## 3.5 ANALYSIS: JOB MODULE

In the Module list located under the toolbar, click Job to open the Job module.

### 3.5.1 CREATE AN ANALYSIS JOB: JOB-1

From the main menu bar, select Job→Manager and the Job Manager window appears.

In the Job Manager, click Create. The Create Job dialog box appears with a list of models in the model database.

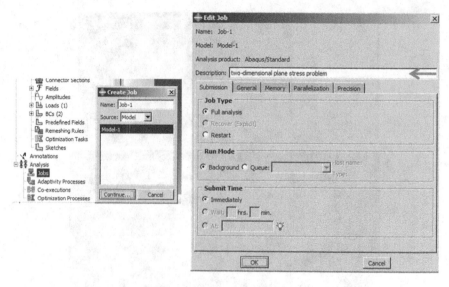

**FIGURE 3.27**   Create a job for analysis.

Name the job and click Continue. The Edit Job dialog box then appears.

Enter a description for the job, type *two-dimensional plane stress problem*.

Check Full analysis and choose to run the job in the background and check to start it immediately, as shown in Figure 3.27. (Note: Job Type is set to Full analysis to combine the data check and analysis phases of the simulation.)

Click OK to accept all other default job settings in the job editor and to close the dialog box.

Expand the tree under Jobs, right click on Job-1. Then, click on Submit (see Figure 3.28).

When the model simulation is completed, the analysis can then be executed. Since errors in the model originating from incorrect or missing data cannot be identified initially, therefore, the user should perform a data check analysis before running the analysis. To run a data check analysis:

First, make sure that the Job Type is set to Data check. From the buttons on the right edge of the Job Manager, click Submit to submit the job for analysis or the user can click on the Data Check at the right edge of the Job Manager for checking purposes. The model will then be on Check Running status.

Once the Abaqus/CAE has updated that the model is in Check Completed status, the user can click on Continue having completed the analysis. The user can also indicate the job's status after job submission. The Status column for the overhead hoist problem shows one of the following:

None while the analysis input file is being generated.

- Submitted while the job is being submitted for analysis.
- Running while Abaqus/CAE is analyzing the model.

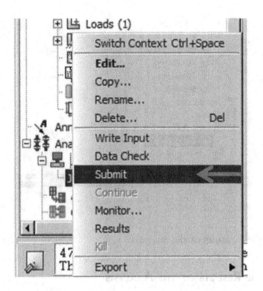

**FIGURE 3.28** Submit Job-1 for analysis.

- Completed when the analysis is completed, and the output has been written to the output database, and the user can click on Results to proceed to the visualization module. Aborted if Abaqus/CAE detects a problem with the input file or the analysis, and consequently, the analysis is aborted. In addition, Abaqus/CAE reports the problem in the message area.

### 3.5.2 MONITOR SOLUTION IN PROGRESS

From the buttons shown on the right edge of the Job Manager, click on Monitor to open the job monitor dialog box once the job is submitted.

The top half of the dialog box displays the information available in the status (*.sta) file that Abaqus creates for the analysis.

If the status on top of the Job Monitor window is shown as Completed, then this job is free of errors and has been executed properly. Then the analysis results can be checked.

## 3.6 VISUALIZATION MODULE

Graphical post-processing is important given the vast volume of data created during a simulation. For any realistic model, it is impractical to interpret the results in the tabular form of the data file. Abaqus/Viewer allows the user to view results graphically using a variety of methods, including deformed shape plots, contour plots, vector plots, animations, and X–Y plots.

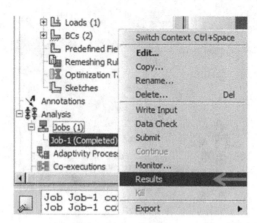

**FIGURE 3.29**  View results of the analysis.

### 3.6.1  VIEW THE RESULTS OF THE ANALYSIS

Once the job is completed, the user can then view the results of the analysis at the Visualization module. From the buttons shown on the right edge of the Job Manager, click Results. Abaqus/CAE loads the Visualization module, opens the

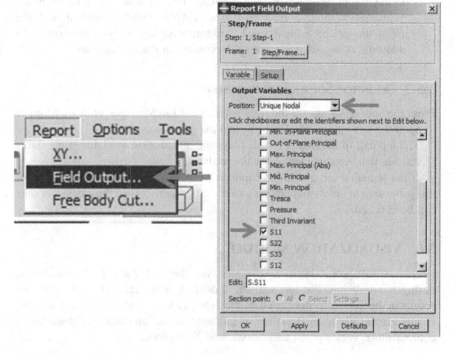

**FIGURE 3.30**  Select the desired results to print to a report.

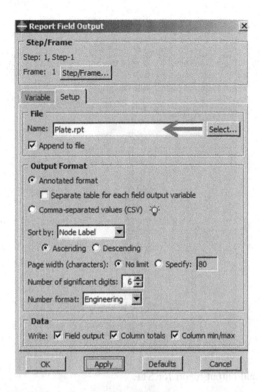

**FIGURE 3.31**  Choosing a directory and the file name to which to write the report.

output database created by the job, and displays a fast plot of the model, as shown in Figure 3.29. A fast plot is a basic representation of the undeformed shape of the model. Alternatively, the visualization option can be clicked in the Module list located under the toolbar; select File→Open and select *Plate.odb* from the list of available output database files, and then click OK.

### 3.6.2  VISUALIZATION/RESULTS MODULE

Abaqus/CAE allows the user to write data to a text file (*.rpt) in a tabular format. This feature is found to be very convenient in writing tabular output to the data (*.dat) file. It is also very useful, especially in writing a report. In this problem, the user will generate a report containing the element stress S11 and the element strain energies. To generate field data reports:

From the main menu bar, select Report→Field Output.

To create a text file containing the values of S11, click on Report in the main menu bar and then click on Field output. In the Report Field Output dialog box, for Position, select Unique nodal, and then check S11 under the expanded list of S: Stress components (see Figure 3.30).

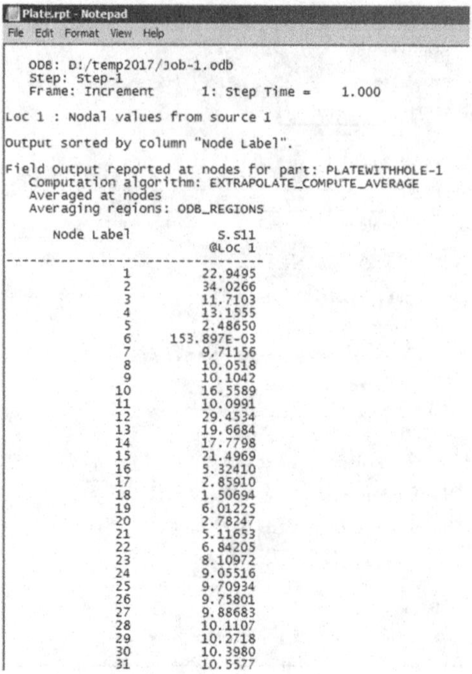

```
Plate.rpt - Notepad
File  Edit  Format  View  Help

  ODB: D:/temp2017/Job-1.odb
  Step: Step-1
  Frame: Increment      1: Step Time =    1.000

Loc 1 : Nodal values from source 1

Output sorted by column "Node Label".

Field Output reported at nodes for part: PLATEWITHHOLE-1
  Computation algorithm: EXTRAPOLATE_COMPUTE_AVERAGE
  Averaged at nodes
  Averaging regions: ODB_REGIONS

        Node Label              S.S11
                               @LOC 1
  -------------------------------------
                 1            22.9495
                 2            34.0266
                 3            11.7103
                 4            13.1555
                 5             2.48650
                 6          153.897E-03
                 7             9.71156
                 8            10.0518
                 9            10.1042
                10            16.5589
                11            10.0991
                12            29.4534
                13            19.6684
                14            17.7798
                15            21.4969
                16             5.32410
                17             2.85910
                18             1.50694
                19             6.01225
                20             2.78247
                21             5.11653
                22             6.84205
                23             8.10972
                24             9.05516
                25             9.70934
                26             9.75801
                27             9.88683
                28            10.1107
                29            10.2718
                30            10.3980
                31            10.5577
```

**FIGURE 3.32**  Field output report for S11.

In the Setup tabbed page, name the report as *Plate.rpt*. This file can also be saved in the user's working directory by clicking on the Select button. In the Data region at the bottom of the page, toggle off the Column totals (the column total is useful when the user wants to compute, for example, the total strain energy of the entire model).

Click Apply (see Figure 3.31). The stress values S11 are appended to the report file (see Figure 3.32).

Abaqus/CAE allows the user to plot the graph of each output.

Select the Create XY Data tool in the visualization toolbox to plot the S11 graph. The Create XY Data dialog box appears (see Figure 3.33).

Check the ODB field output in the list of sources. Then click on Continue. XY Data from the ODB Field Output dialog box appears. In the Variables tabbed page, scroll through the list of positions and select the UniqueNodal.

Expand the S: Stress Components and check S11 (see Figure 3.34).

In the Elements/Nodes tabbed page, click Pick from the viewport and then click Edit Selection (see Figure 3.35).

In the next step, the user is allowed to pick any element available in the viewport. Once the element is selected, click on Done and it shows that one element is selected.

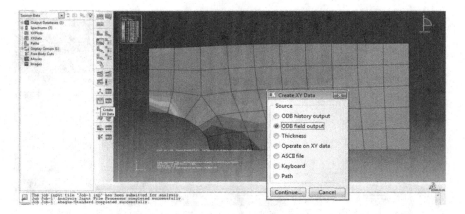

**FIGURE 3.33**   Create/operate on XY data.

**FIGURE 3.34**   Select the desired results for graph plotting.

Click on Plot. The graph of Stress vs. Time at that selected element is plotted in the viewport (see Figure 3.36).

The S11 stress contour can be observed by clicking on the icon Plot Contours, on the Deformed Shape (see Figure 3.37).

Note: If the user cannot read the S11 values in the legend block on the main menu, click on View port Annotation Options. Under Legend, click on Set font and enter a larger font (see Figure 3.38).

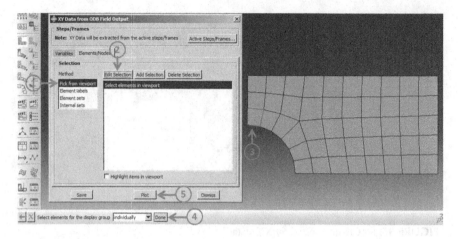

**FIGURE 3.35** Selecting the element from the viewport.

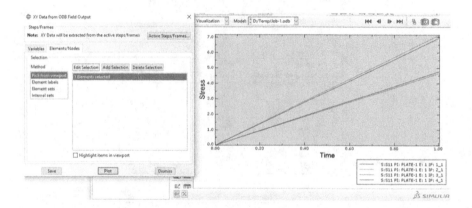

**FIGURE 3.36** Stress vs. time plot.

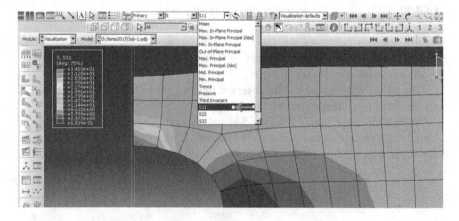

**FIGURE 3.37** View the deformed shape of the plate.

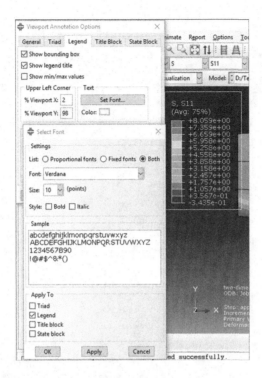

**FIGURE 3.38** Viewport annotations options.

The red-shaded contour appears on the plate, indicating that it has experienced the highest stress when the load is applied. The stress is in a positive value if the region experiences tension. On the other hand, when the region experiences compression, it shows a negative value of stress.

The stress increases as it comes nearer to the predefined constraint.

# 4 Evaluate the Capacity of the Steel Beam-Column Connection through Pushover Analysis

## 4.1 INTRODUCTION

This chapter is discussing about performing a nonlinear static pushover analysis for conventional steel beam-column connections using the Abaqus/Standard the finite element program. The analysis is performed on the three-dimensional model to study the efficiency of the configuration used in conventional steel beam-column connections. In this example, the beam is subjected to two different applied loads, which are the concentrated load and the uniform distributed load. Two case scenarios have also been used to evaluate the stability of the connections while the distributed load increased by a factor of 2. The beam is connected to the column by a full penetration weld.

Then, results have been investigated in terms of base shear vs. displacement, displacement vs. time, and base shear vs. time. The numerical result indicated that joint capacity for both cases showed the same behavior in the elastic range, whereas, its behavior changed in the plastic range.

## 4.2 PROBLEM DESCRIPTION

The model consists of a beam and a column which is connected by full penetration weld connection as illustrated in Figure 4.1.

### 4.2.1 GEOMETRIC PROPERTIES

The beam and the column are universal beam section. The beam and the column section are UB 254 × 102 × 25 and UB 356 × 127 × 33, respectively. Table 4.1 shows the section properties of the beam and column. The joint has been strengthened by a plate of 3mm thickness, which is used in two positions, as illustrated in Figure 4.2.

DOI: 10.1201/9781003213369-4

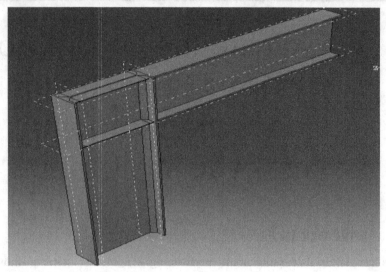

**FIGURE 4.1**  3D Model.

**TABLE 4.1**
**Geometric properties**

| Designation | Depth of section (mm) | Width of section (mm) | Thickness of | | Root Radius (mm) | Depth between Fillets (mm) |
|---|---|---|---|---|---|---|
| | | | Web (mm) | Flange (mm) | | |
| | h | b | s | t | r | d |
| 254 × 102 × 25 | 257.2 | 101.9 | 6 | 8.4 | 7.6 | 225.2 |
| 356 × 127 × 33 | 349 | 125.4 | 6 | 8.5 | 10.2 | 311.6 |

### 4.2.2 MATERIAL PROPERTIES

Table 4.2 shows the material properties of the steel section for both the beam and column.

## 4.3 OBJECTIVES

1. To investigate the efficiency of the configuration used in conventional steel beam-column connections.
2. To investigate the joint capacity of the column-beam connections, which are linked together by weld connections.
3. To study the stability of steel beam-column connections under two types of loadings applied.

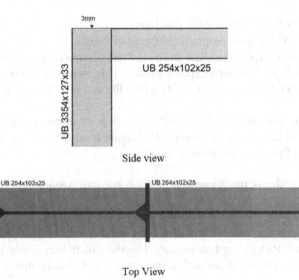

FIGURE 4.2 Cross Section of the beam-column connection.

**TABLE 4.2**
**Material Properties**

| Properties | Quantity |
| --- | --- |
| Density | 7.85 E9 kg/m3 |
| Young Modulus | 210,000 MPa |
| Poisson ratio | 0.3 |
| Yield Stress | 370 MPa |
| Failure Stress | 460 MPa |
| Plastic strain at yield | 0 |
| Plastic strain at Failure | 0.25 |

## 4.4 MODELING

### 4.4.1 PART MODULE

The part module is used to create various parts of the model. In this case, the model is divided into six parts: Column, Beam, Segment left-top, Segment right-top, segment left-middle, and segment right-middle. To create a 3-D geometry, first, a 2-D profile should be created and then manipulating it to obtain the solid geometry. The following subsections show a step-by-step procedure on creating each of these parts.

### 4.4.1.1   Create a new model database

Start Abaqus/CAE from programs in the Start menu.

Select Create Model Database from the Start Session dialog box that appears.

Click on With Standard/Explicit Model. This step allows the user to start modeling whereby the user can create a new file and save it under any name in a new folder.

When the Part module has finished loading, it displays the Part module toolbox in the left side of the Abaqus/CAE main window. Each module displays its own set of tools in the module toolbox.

Column:

Figure 4.3 shows the dimension of the column in Abaqus.

### 4.4.1.2   Create a new model database and a new part

From the menu bar, select Part→Create in order to create a new part.

The Create Part dialog box appears. Use the Create Part dialog box to name the part and to choose its modeling space, type, and base feature and to set the

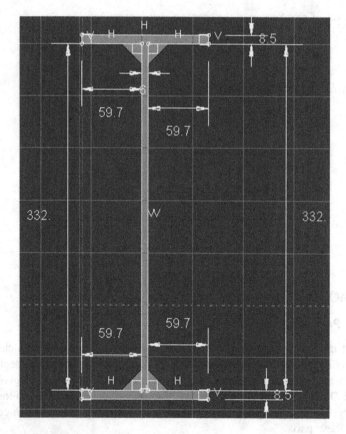

**FIGURE 4.3**   Column Cross-section.

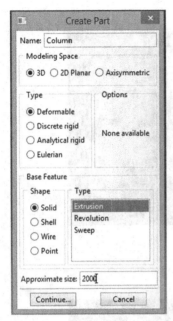

**FIGURE 4.4** Create Column.

approximate size. The name of the part may be edited once it has been created, but the modeling space, type, or base feature cannot be changed.

Name the part *Column*. Next, choose three-dimensional, deformable body type and Solid and Extrusion as the base feature (see Figure 4.4).

Enter an approximate size of 2000. The value entered in the approximate size text field at the bottom of the dialog box sets the approximate size of the new part.

Click Continue to exit the Create Part dialog box.

### 4.4.1.3 Section of the column with dimension

Use the Create Lines: connected tool located in the upper right corner of the Sketcher toolbox to begin drawing the geometry of the column. The user can choose a starting corner for the column at the viewport or enter the X and Y coordinates. Next, create a line with known coordinates, or alternatively, the user can define the dimension of the geometry by clicking on the add dimension tool. Once finished sketching the section for the dimension, right click and click on Cancel Procedure to exit the sketcher as shown in Figure 4.5.

Click on Done at the prompt area once finished, and it will be displayed, as shown in Figure 4.6. The user needs to define the depth for the extrusion of the column. Once the depth of the column has been defined, it will be displayed, as shown in Figure 4.7.

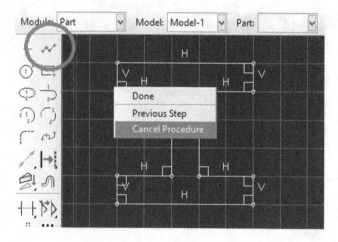

**FIGURE 4.5**   Draw column section.

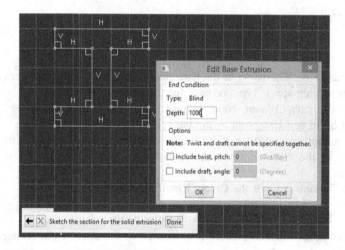

**FIGURE 4.6**   Edit Base Extrusion.

### 4.4.1.4   Create a chamfer

Create a chamfer by selecting the Create Round or Fillet from the Sketcher toolbox. The user needs to select the edges to round/fillet, then define the chamfer radius, as shown in Figure 4.8 and Figure 4.9.

### 4.4.1.5   Beam

The steps for the beam formation are similar to that for the column, but it should be drawn with different dimensions, as shown in Figure 4.10.

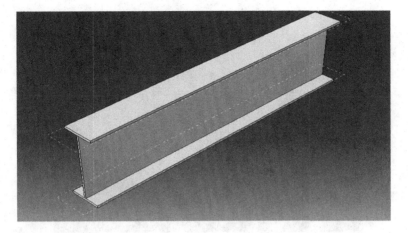

**FIGURE 4.7**   Column.

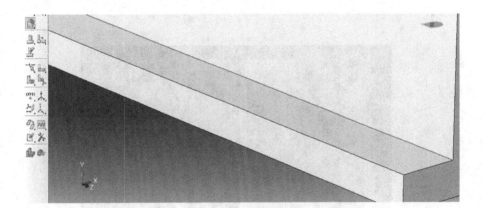

**FIGURE 4.8**   Chamfer Creation.

### 4.4.1.6   Segments

The segments are used for covering the top of the column on both sides. The steps for creating the segments are the same as for the previous part instances. Figure 4.11 shows the dimension of the segments. The plate should be fit inside the column, as shown in Figure 4.12.

### 4.4.2   PROPERTY MODULE

In this module, the material properties for the analysis should be defined by assigning those properties to the available parts.

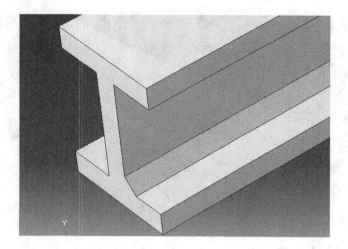

**FIGURE 4.9**   Chamfer Done.

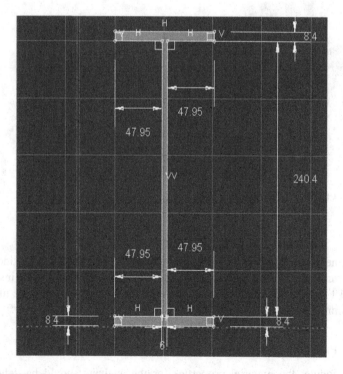

**FIGURE 4.10**   Beam Section Details.

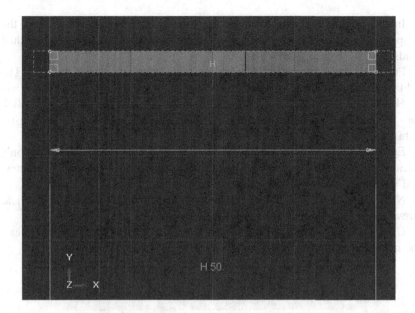

**FIGURE 4.11** Segments Dimensions.

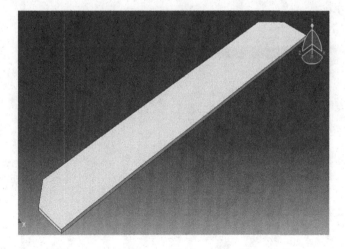

**FIGURE 4.12** 3D view of Segments.

### 4.4.2.1 Material Properties

The Property module is used to define the properties of various materials used in the model. Sections are then created, and materials assigned to each section. The property of this model is made of steel as discussed earlier and assumed to be linear elastic with Young's modulus of 210,000 N/mm$^2$ and Poisson's ratio of 0.3. To define a material, follow the steps below:

In the Module list located under the toolbar, select Property to open the Property module. The cursor changes to an hourglass while the Property module loads.

From the menu bar, select Material→Create, to create a new material. The Edit Material dialog box appears.

Name the material *Steel*. From the material editor's menu bar, click on Mechanical→Elasticity→Elastic. Abaqus/CAE displays the Elastic data form.

Enter the value of 210,000 MPa for Young's Modulus and 0.3 for Poisson's Ratio in the respective cells. Use the [Tab] button or move the cursor to a new cell and click to move between cells.

Next, click on Mechanical → Plasticity → Plastic. Enter the values for yield stress and plastic strain, as shown in Figure 4.13.

Click General → Density. Next, enter the value 7.85E-009. Click OK to exit the material editor.

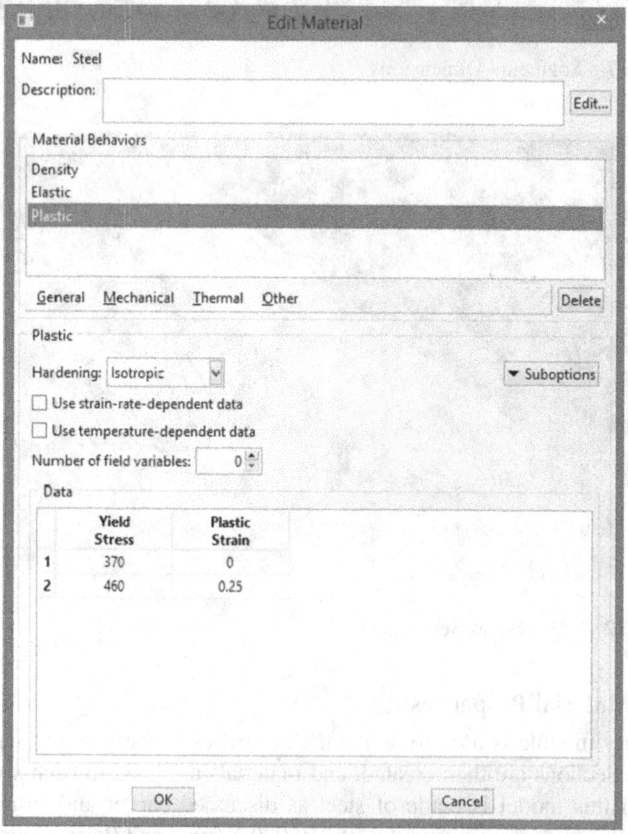

**FIGURE 4.13** Properties for steel.

### 4.4.2.2 Section properties

The section properties of a model are defined by creating sections in the Property module. After the section is created, one of the following two methods to assign the section to the part can be used.

Select the region from the part and assign the section to the selected region, or use the Set toolset to create a homogeneous set containing the region and assign the section to the set.

To define a beam section:

From the main menu bar, select Section→Create. The Create Section dialog box appears.

In the Create Section dialog box:

Name the section *Beam*.

In the Category list, select Solid. In the Type list, select Homogeneous (see Figure 4.14). Click Continue. The Edit Section dialog box appears.

In the Edit Section dialog box:

Click the arrow next to the Material text box and scroll through the Materials to see a list of available materials and to select the required material. To define each material, it is necessary to do the same process of Create Section and then to determine the material by clicking on the *Continue* button.

In the Plane stress/strain thickness field, accept default. Then, Click OK.

### 4.4.2.3 Section Assignment

Next, assign the defined section to the corresponding part. The Assign menu is used in the Property module to assign the section *Beam* to the beam. To assign the section to the beam:

From the main menu bar, select Assign→Section. Abaqus/CAE displays prompts in the prompt area to guide the user through the procedure.

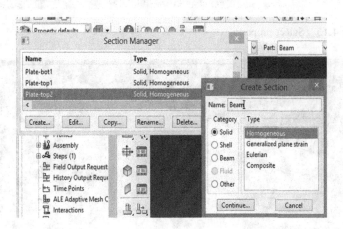

**FIGURE 4.14** Create section property.

Alternatively, expand the menu under the Beam and double click on the Section Assignments.

Select the entire part as the region to which the section will be applied.

Click and hold down the left mouse button and at the upper left corner of the viewport:

Drag the mouse pointer to create a box around the beam.

Release the left mouse button. Abaqus/CAE highlights the entire beam.

Right-Click on the viewport or click Done in the prompt area to accept the selected geometry. The Assign Section dialog box appears.

In Section, scroll to Beam and click OK as shown in Figure 4.15. The Part changes color to green once the section is assigned, as shown in and Figure 4.16.

Different Assigned color:

While the section has a different color, it means that different cases exist to identify the status of the section. When the section is red in color, it means that the section was deleted or has no properties, as shown in Figure 4.17. In other

**FIGURE 4.15**  Select property.

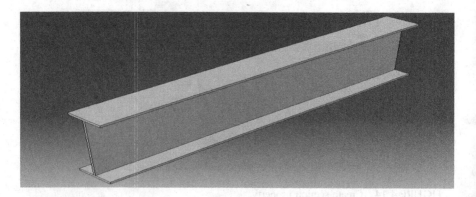

**FIGURE 4.16**  Assigned section.

**FIGURE 4.17** No assigned section.

words, the green color represents the assigned section, as shown in Figure 4.18, and the yellow color shows that the section has been assigned two times, as shown in Figure 4.19.

### 4.4.3 ASSEMBLY MODULE

In this module, all the parts created earlier can be put together (assembly) to achieve the required model. The constraints and loads can be applied to the model once all the individual part instances are assembled.

#### 4.4.3.1 Assemble part instances into the model

In the Module list located under the toolbar, click Assembly to open the Assembly module. The cursor changes to an hourglass while the Assembly module loads. From the main menu bar, select Instance→Create. The Create Instance dialog box appears.

In the opened window under the Instance Type box, choose Dependent (mesh on the part).

The first step, the instance part, is to bring all parts together which have been previously created.

In the dialog box, select Beam and click OK. In the dialog box, select Column and click OK.

Click on Translate Instance to move the elements to the required position.

Select one node from the top node of the column.

Select the center of the top flange of the beam, so that they fit together, as shown in Figure 4.20. Sometimes the translation cannot be performed correctly given the parts need to rotate. For this reason, one of the parts either the column or the beam must rotate until it fits the other part.

Click on Rotate Instance to rotate any part about any axis to suit the right direction.

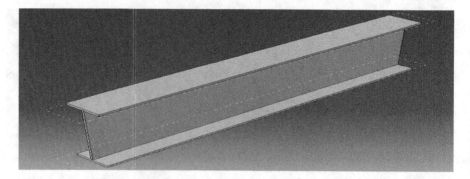

**FIGURE 4.18** Assigned Section.

**FIGURE 4.19** Duplicated assigned section.

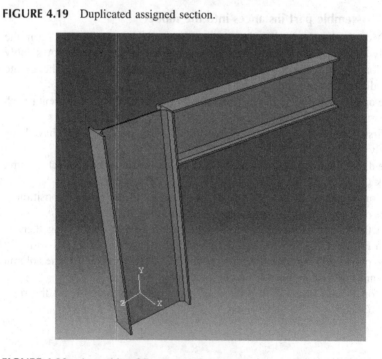

**FIGURE 4.20** Assembly of Beam and Column.

Repeat the above steps for Segment Top Left and select the proper point on the segment that can be fitted to the column to complete the model by using the Translate Instance tool as shown in Figure 4.21.

### 4.4.4 Step Module

After finishing the assembly section, the configuration of the analysis should next be defined. In this simulation analysis, we are interested in identifying the static response of the beam-column connection structure. Abaqus/CAE generates the initial step automatically. However, the rest of the analysis step should be carried out by the user as well as the requested output for any steps in the analysis. There are two kinds of analysis steps in Abaqus: general analysis steps, which can be used to analyze the linear or nonlinear response, and the linear perturbation steps, which can be used only to analyze linear problems. Only general analysis steps are available in Abaqus/Explicit. Next, create a dynamic, nonlinear perturbation step that follows the initial step of the analysis. This module is used to perform many tasks, mainly to create analysis steps and to specify output requests.

#### 4.4.4.1   Create an analysis step: *Pushover*

After completing the assembly of model, the Step module needed to done to configure the required analysis. In this simulation analysis, the aim is to evaluate the static response of the structure under loads. This is a single event, so only a single analysis step is needed for the simulation. Thus, the analysis will consist of two steps:

FIGURE 4.21   Segment assembling.

- The initial step, in which boundary conditions that constrain the end of the plate are applied
- The analysis step, in which a distributed load is at the top surface of the beam element.

Abaqus/CAE generates the initial step automatically, but a Step module has to be operated by the user to create the analysis step. The Step module also allows the user to request output (field output and history output) for each step in the analysis.

In the Module list located under the toolbar, click Step to open the Step module.

From the main menu bar, select Step→Create, to create a step. The Create Step dialog box appears with a list of all general procedures and a default step named *Step-1*.

Change the step name to *Pushover*.

Select general as the Procedure type.

Scroll through the available list, select Static, General, and click on Continue. (see Figure 4.22)

Then, the Edit Step dialog box appears.

In the Basic tab, enter the time period with a value of 10 seconds, and the rest is selected by default.

Click the Incrementation tab and change the increment size, as shown in Figure 4.23.

Click the Other tab to see its contents; accept the default values provided for the step.

Click OK to create the step and to exit the Edit Step dialog box.

At the amplitude field, a tabular form of amplitude-time step is defined as shown in Figure 4.24. This option allows arbitrary time variations of load, displacement, and other prescribed variable magnitudes to be given throughout a step.

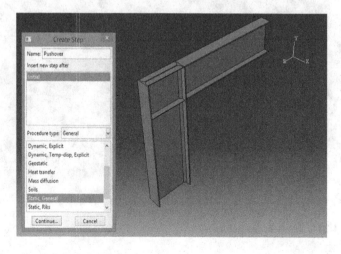

**FIGURE 4.22**   Create Step.

Name: Step-2
Type: Static, General

| Basic | Incrementation | Other |

Type: ● Automatic ○ Fixed

Maximum number of increments: 100

|             | Initial | Minimum | Maximum |
|-------------|---------|---------|---------|
| Increment size: | 1   | 1E-005  | 1       |

**FIGURE 4.23**   Incrementation value.

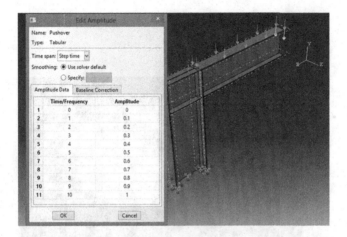

**FIGURE 4.24**   Amplitude creation.

## 4.4.4.2   Create amplitude

Under the model tree, double-click on the Amplitude. The Create Amplitude dialog box appears. Name the amplitude Pushover. Select the Tabular type of amplitude and click Continue. Input the incremental of the amplitude data, as shown in Figure 4.24.

Click OK to exit from the dialog box.

To create a homogeneous set, use the Set toolset. From the main menu bar, select Tool→Set→Create, and the Create Set dialog box appears.

In the Create Set dialog box:

Name the set *Displacement* (see Figure 4.25).

Select Node as the type of set and click Continue. At the viewport, select one node on the edge of the cantilever (see Figure 4.26). Click Done to complete the node assignment to the set.

Repeat the same procedure repeat by selecting the nodes at the end column. (see Figure 4.27)

To obtain the required outputs at the visualization module, it is necessary to request the history outputs in Step Module. The steps are as follow:

**FIGURE 4.25** Create Set.

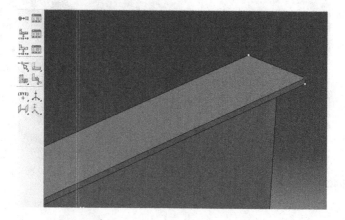

**FIGURE 4.26** Select tip nodes.

Click on History Output Requests Manager to create History Output Requests. The Create History dialog box appears.

Name the history output *Displacement* and click Continue. (see Figure 4.28)

The Edit History Output Request dialog box appears. At the domain, select Set. Click the field next to the domain and scroll through the Set to see a list of the available sets and to select the required set as shown in Figure 4.29.

Expand the Displacement/Velocity/Acceleration and toggle on U2 under U, Translations, and rotations (see Figure 4.29). Click OK to exit from the dialog box.

Repeat the same procedure for the Base Shear at the end column nodes as a reference surface for the base shear result by requesting the reaction force at X-direction, RF1 under the Forces/Reactions.

Note: The Software provides initial analyses, which are Stress, Strain, Reactions, and displacements for the whole structure. This result is considered

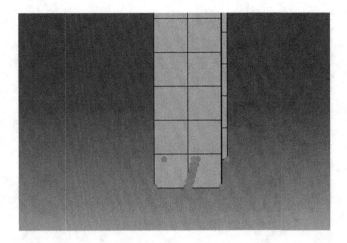

**FIGURE 4.27** Select base nodes.

**FIGURE 4.28** Create history output.

primary as given by the software. The user can still specify the stress and strain and other parameters by clicking on the create field output icon.

### 4.4.5 INTERACTIONS MODULE

This module is used to define various interactions within the model or interactions between regions of the model and its surroundings. The interactions can be mechanical or/and thermal. Analysis constraints can also be applied between two similar and different materials such as Tie connection, Rigid, embedded region, shell to solid coupling, etc. The interaction used in this example is a weld connection between the beam and column and also between the column and segments. To demonstrate the condition of the weld connection, the tie constraints in Abaqus will be implemented in this model.

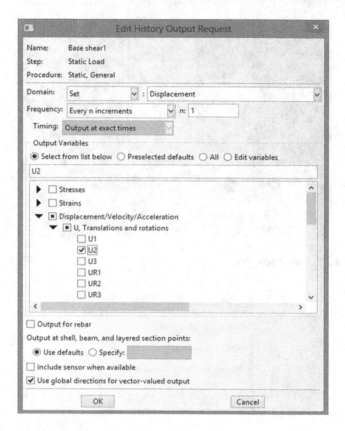

**FIGURE 4.29**  Edit History Output Request.

### 4.4.5.1  Tie Constraint

Structural welding is a heating/fusing process in which enables the parts to be connected [joined], with supplementary molten metal at the joint. A relatively small depth of material will become molten, and upon cooling, the structural steel and weld metal will act as one continuous part where they are joined as displayed in Figure 4.30.

In Abaqus, fully constrained contact behavior can be defined using tie constraints. A tie constraint provides a simple way to bond surfaces together permanently and ties two separate surfaces together so that there is no relative motion between them. Moreover, it is a surface-based constraint using a master-slave formulation in which the constraint prevents slave nodes from separating or sliding relative to the master surface. This type of constraint allows two regions to fuse together even though the meshes created on the surfaces of the regions may be dissimilar. In this example, the tie connection is used to link the beam and column together. Tie rotational degree of freedoms are also used, if applicable, to allow the connection to have a possibility of rotation in a default value.

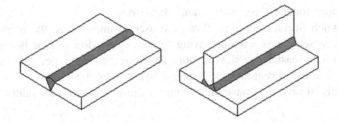

**FIGURE 4.30** 3D Weld connection.

### 4.4.5.2 Tie the surfaces

In the Module list located under the toolbar, click Interaction to open the Interaction module.

From the main menu bar, select Constraint→Create to create a constraint. The Create Constraint dialog box appears with a list of all types of constraints and a default constraint named *Constraint-1*.

Name the constraint as *Constraint-1*

Select Tie as the type of constraint and click on Continue (see Figure 4.31). In the prompt area, choose the master type: Surface.

Select regions for the master surface and then click Done. There are two ways to select the surface:

The user can use the existing surface to define the region. On the right side of the prompt area, click Surfaces and select an existing surface name from the Region Selection dialog box that appears and click Continue.

The user can use the mouse to select a region in the viewport. The main surface of the column is selected as the master surface (see Figure 4.32). In the prompt area, choose the slave type: Surface. The left edge surface of the beam is selected (see Figure 4.33).

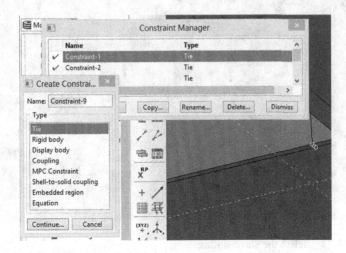

**FIGURE 4.31** Create interaction type.

Select regions for the slave surface and click Done.

The Switch Surfaces option allows the user to interchange the master and slave surface selections without having to start over. However, it is available only when both master and slave regions are from the same type; both surfaces and both node-based regions (see Figure 4.34). Accept the default.

Click OK to save the constraint definition and to exit from the editor.

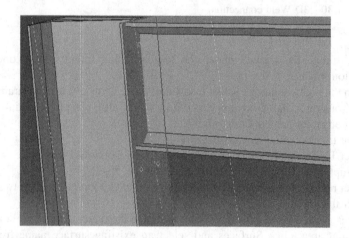

**FIGURE 4.32**   Select the main surface.

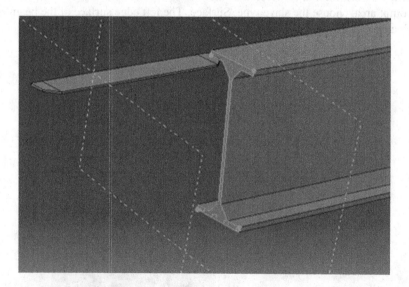

**FIGURE 4.33**   Select the slave surface.

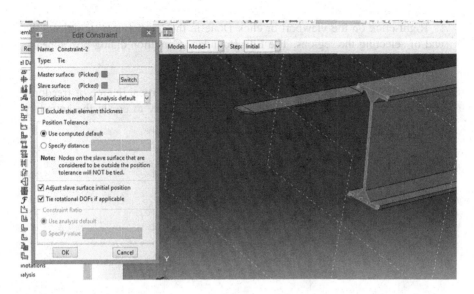

**FIGURE 4.34** Create interaction.

## 4.4.6 BOUNDARY CONDITION

Using this command, the user can create the loading that needs to be assigned to the elements or parts of the structure. The prescribed conditions, such as loads and boundary conditions, are step dependent, which means that the user needs to specify the steps in which they become active. In the next step, that the steps in the analysis have been defined, the Load module can be used to define the prescribed conditions.

Apply boundary conditions to the plate:

In the Module list located under the toolbar, click Load to open the Load module.

From the main menu bar, select BC→Create. The Create Boundary Condition dialog box appears.

In the Create Boundary Condition dialog box:

Name the boundary condition *Fix End*.

From the list of steps, select Pushover as the step in which the boundary condition will be activated. All the mechanical boundary conditions specified in the Initial step must have zero magnitudes. This condition is enforced automatically by Abaqus/CAE.

In the Category list, accept Mechanical as the default category selection.

In the Types for Selected Step list, select Symmetry/Axisymmetry/Encastre, and click Continue. Abaqus/CAE displays prompts in the prompt area to guide the user through the procedure (see Figure 4.35).

There are several boundary conditions identified by the software such as pinned, fixed, etc.

In the viewport, select the base of the column. This is the region to which the boundary condition will be applied (see Figure 4.36).

Right-click on the viewport or click Done in the prompt area to indicate the end of selecting the regions. The Edit Boundary Condition dialog box appears. In the dialog box:

Fill the dialog box, as demonstrated in Figure 4.37.

Select ENCASTRE, since all the horizontal and vertical degree of freedoms needed to be constrained as a fixed support. Click OK to create the boundary condition and to close the dialog box.

**FIGURE 4.35**  Create a boundary condition.

**FIGURE 4.36**  Select Boundary region.

The boundary conditions at the column base will be illustrated, as shown in Figure 4.38

### 4.4.7 LOAD MODULE

The concentrated load has been applied on the beam edge, and distributed load

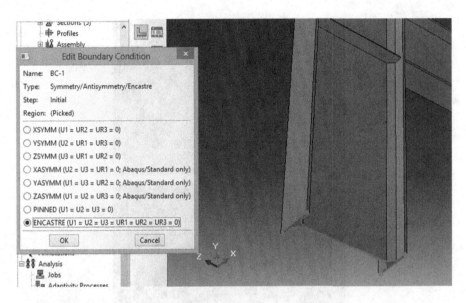

**FIGURE 4.37** Edit boundary condition.

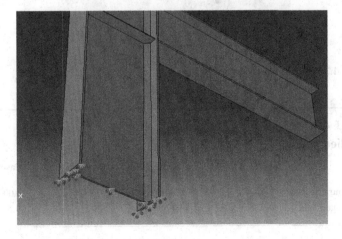

**FIGURE 4.38** Fix End Boundary Condition.

has also been applied on the whole top beam surface (see Figure 4.39). The quantity and type of applied loads can be seen in Table 4.3.

The concentrated load and distributed load are both applied to the model at the same location in order to indicate the effect of both point load and concentrated load.

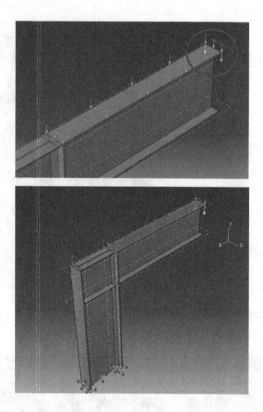

**FIGURE 4.39**   Load Configurations.

**TABLE 4.3**
**Applied Loads**

| Case | Case 1 | Case 2 |
| --- | --- | --- |
| Concentrated load | 20,000 N | 20,000 N |
| Distributed load | 50 N/mm² | 100 N/mm² |

After generating a boundary condition, it is possible to define loads for the model. In this example, two types of loading will be used to indicate the effect of push over loading. The concentrated load is applied in the initial step, which was created in the Step module. To apply the concentrated load to the model, carry out the following steps:

From the menu bar, select Load→Manager.

At the bottom of the Load Manager, click Create. The Create Load dialog box appears.

In the Create Load dialog box:

Name the load *Concentrated.*

Scroll through the available list of steps, select Static Load as the step in which the load will be exerted. In the Category list, accept Mechanical as the default category selection. In the Types for Selected Step list, select Concentrated Force and click Continue (see Figure 4.40).

Abaqus/CAE displays prompts in the prompt area to guide the user through the procedure. The user is asked to select a region to which the load will be applied. As with the boundary conditions, the region to which the load will be applied can be selected either directly in the viewport or from a list of existing sets. As before, select the region directly in the viewport. In the viewport, select the points on the cantilever edge of the beam (see Figure 4.41).

Click on the viewport or click Done in the prompt area to finish selecting the regions. The Edit Load EditLoad dialog box appears. In the dialog box: Enter a magnitude of -20,000 N as a vertical concentrated load. The negative sign in front of the value indicates the load applied is a downward force as gravitational force is acting downwards (see Figure 4.42). Click OK to create the load and to close the dialog box. Figure 4.43

The concentrated load will appear on the cantilever edge of the beam, as shown in Figure 4.43.

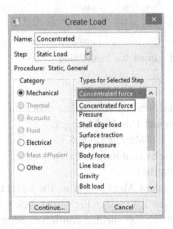

**FIGURE 4.40**   Create Loading.

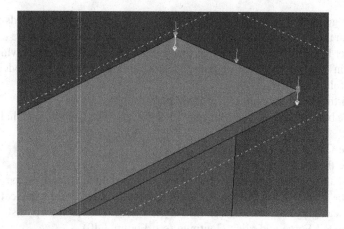

**FIGURE 4.41**   Select points.

| Edit Load | × |
|---|---|
| Name: Concentrated | |
| Type: Concentrated force | |
| Step: Static Load (Static, General) | |
| Region: (Picked) Edit Region... | |
| CSYS: (Global) Edit... ⟂ Create... | |
| Distribution: Uniform ∨ Create... | |
| CF1: 0 | |
| CF2: ½0000 | |
| CF3: 0 | |
| Amplitude: (Ramp) ∨ Create... | |
| ☐ Follow nodal rotation | |
| **Note:** Force will be applied per node. | |
| OK Cancel | |

**FIGURE 4.42**   Input the load Value.

For uniform distributed loading follow the procedure below:

From the menu bar, select Load→Manager.

At the bottom of the Load Manager, click Create. The Create Load dialog box appears. Scroll through the available list of steps, select *Static Load* as the step in which the load will be exerted.

In the Category list, accept Mechanical as the default category selection. In the Types for Selected Step list, select Pressure.Click Continue.

In the viewport, select the top surface cantilever edge of the beam where the load will be applied.

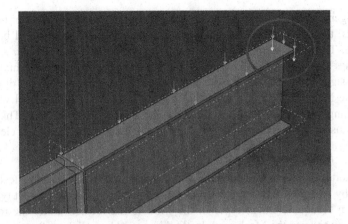

**FIGURE 4.43**   The Concentrated Load.

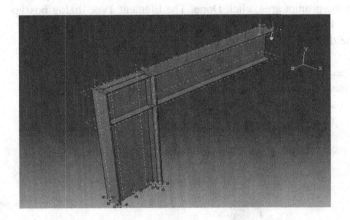

**FIGURE 4.44**   The Distributed Load Case 1.

Click on the viewport or click Done in the prompt area to finish selecting the regions. The Edit Load dialog box appears.

In the dialog box:

Enter a magnitude of $50N/m^2$ as a distributed load. Click OK to create the load and to close the dialog box.

The uniform distributed load will appear on the top surface cantilever edge of the beam, as shown in Figure 4.44.

### 4.4.8   Mesh Module

This is one of the most important modules given the accuracy of the results will rely on the number of mesh elements and mesh element size. This module can be used to generate meshes as well as to verify them. The larger the number of mesh elements, the smaller the mesh element size, and thus, the results will be more accurate. The finite element mesh should be generated in this module. Abaqus/CAE uses a number of different meshing techniques. The default meshing technique assigned to the model is indicated by the color of the model that is displayed when the Mesh module is opened. If Abaqus/CAE displays the model with the color orange, it cannot be meshed without the assistance of the user. This command is used to mesh the whole structure to small and equal parts and elements.

### 4.4.8.1   Mesh: Seed the part

Mesh module is used to generate the finite element mesh. The meshing technique is used by Abaqus/CAE to create the mesh, element shape, and element type. To assign an Abaqus element type: In the Module list located under the toolbar, click Mesh to open the Mesh module. At the context bar, click on Part, to unclick the assembly. From the main menu bar, select Mesh→Element Type.

In the viewport, select the entire frame as the region to be assigned an element type. In the prompt area, click Done. The Element Type dialog box appears, as shown in Figure 4.45

In the dialog box, select the following:

- Standard as the Element Library selection (the default).
- Linear as the Geometric Order (the default).
- 3D stress as the Family of elements.

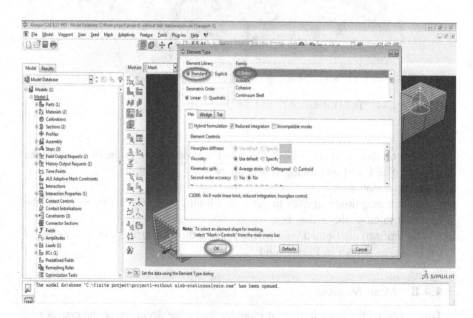

**FIGURE 4.45**   Selecting the Element Type.

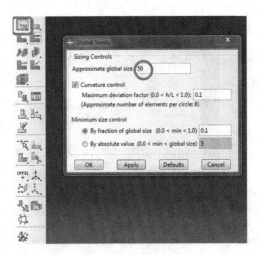

**FIGURE 4.46**   Assign the approximate global size for mesh elements.

The mesh can be created in this step. Meshing is basically a two-stage operation. First, seeding the edges of the part instance, followed secondly by meshing the part instance. Select the number of seeds based on the desired element size or on the number of elements that are required along an edge, and Abaqus/CAE places the nodes of the mesh at the seeds whenever possible. For manual meshing, click on Seed Edges. Then, select the edges that the user wants to mesh. Next, click Mesh Part Instance.

Seed and mesh the model:

From the main menu bar, select Seed→Part to seed the part instance.

Alternatively, select the Seed Part on the upper left corner of the meshing toolbox. The Global Seeds dialog box will appear, as shown in Figure 4.46.

Type the appropriate value for the approximate global size of the mesh elements. Click OK to accept the seeding. Select the Assign Mesh Controls tool at the meshing toolbox, and the Mesh Controls dialog box appears, as shown in Figure 4.47.

From the main menu bar, select Mesh→Instance to mesh the part instance or select the Mesh Part Instance at the upper left corner of the meshing toolbox. Select the part instances to be meshed. Once finished selecting, click Yes in the prompt area to confirm the mesh of the part instance. Once meshed, the plate changes color to blue. The meshed geometry is shown in Figure 4.48.

Abaqus/CAE highlights any elements that fail the mesh quality tests and displays the number of elements tested along with the number of errors and warnings in the message area.

The user can verify the model meshing by clicking on the Verify Mesh icon, as shown in Figure 4.49, and a message will appear in the text messages at the bottom of the screen (see Figure 4.50).

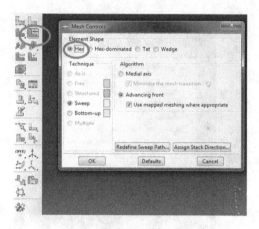

**FIGURE 4.47**   Choose the element shape and meshing technique option.

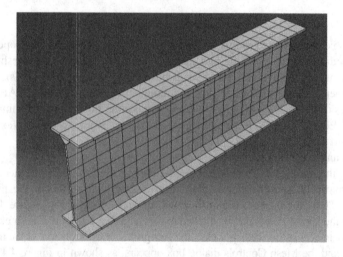

**FIGURE 4.48**   Meshing Done.

The segment has an irregular shape at both edges, which might cause some meshing errors, as shown in Figure 4.51. Therefore, in order to avoid meshing errors, a datum plane is used. A datum plane is a plane that is used to separate the irregular shape from the main body and doing meshing for the parts separately, as shown in Figure 4.52.

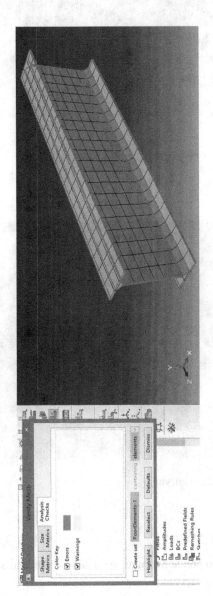

**FIGURE 4.49** Verify Mesh.

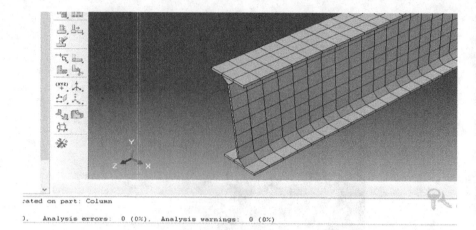

:ated on part: Column

). Analysis errors:  0 (0%). Analysis warnings:  0 (0%)

**FIGURE 4.50** Meshing Verify.

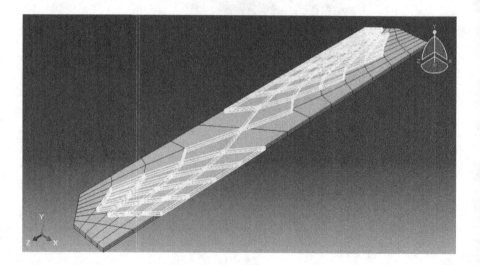

**FIGURE 4.51** Errors given after verifying.

## 4.5 ANALYSIS: JOB MODULE

A job module can be used to create and manage analysis jobs and submit them for analysis.

### 4.5.1 CREATE AN ANALYSIS JOB: JOB-1

From the main menu bar, select Job→Manager and the Job Manager window appears. When finished defining the job, the Job Manager will display a list of

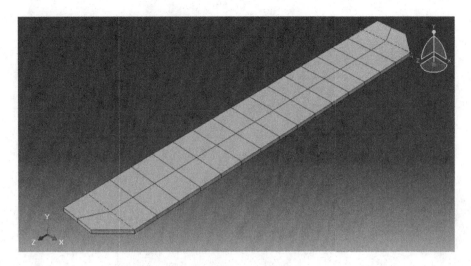

**FIGURE 4.52** Rearranging meshing by the datum plane.

**FIGURE 4.53** Create job.

jobs, the model associated with each job, the type of analysis, and the status of the job. In the Job Manager, click Create. The Create Job dialog box appears with a list of models in the model database.

Name the job Pushover-1 and click Continue (see Figure 4.53).

The Edit Job dialog box appears. In the Description field, type Pushover. In the Submission tabbed page, select Full Analysis as the Job Type (see Figure 4.54).

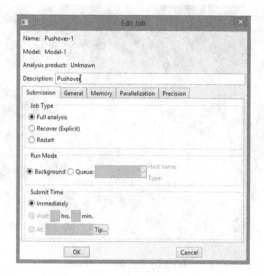

**FIGURE 4.54**  Create job by default.

**FIGURE 4.55**  Job submission.

Click OK to accept all other default job settings in the job editor and to close the dialog box. Click on Submit to start checking the input file and running the analysis (see Figure 4.55).

When the model simulation is finished, analysis can then be executed. Since errors in the model originating from incorrect or missing data cannot be identified initially, the user should perform a data check analysis before running the analysis. To run a data check analysis:

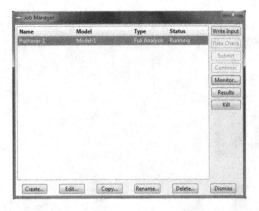

**FIGURE 4.56**   Running the job analysis.

Make sure that the Job Type is set to Data check. From the buttons on the right edge of the Job Manager, click Submit to submit the job for analysis or the user can click on the Data Check at the right edge of the job manager for checking purposes. The model will then be on Check Running status.

After job submission, the information in the Status column is updated to indicate the job's status. The Status column for the overhead hoist problem shows one of the following:

- None while the analysis input file is being generated.
- Submitted while the job is being submitted for analysis.
- Running while Abaqus analyses the model. (see Figure 4.56).
- Completed when the analysis is done, and the output has been written to the output database, and the user can click on the Results to proceed to the visualization module.
- Aborted if Abaqus/CAE finds a problem with the input file or the analysis and aborts the analysis. In addition, Abaqus/CAE reports the problem in the message area.

### 4.5.2   MONITOR SOLUTION IN PROGRESS

Click Monitor to observe errors/warnings. From the buttons on the right edge of the Job Manager, click Monitor to open the job monitor dialog box once the job is submitted.The top half of the dialog box displays the information available in the status (*.sta) file that Abaqus creates for the analysis.If the status on top of the Job Monitor window is shown as Completed, then this job is free of errors and was executed properly. In the next step, the results can be observed.

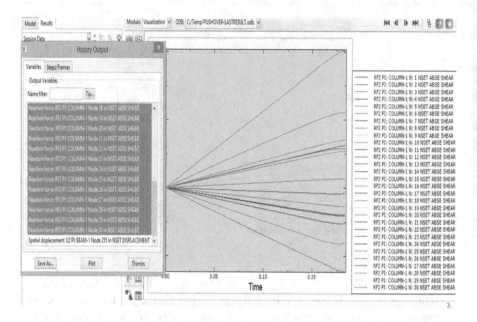

**FIGURE 4.57**   Graphs of Base shear vs. Time.

## 4.6   VISUALIZATION MODULE

Graphical post-processing is important given the vast volume of data created during a simulation. For any realistic model, it is impractical to try and interpret the results in the tabular form of the data file. Abaqus/Viewer allows the user to view the results graphically using a variety of methods, including deformed shape plots, contour plots, vector plots, animations, and X–Y plots. Here, the model can be viewed, and various plots generated. The result of stress and displacement in the main menu bar is requested.

### 4.6.1   VIEW THE RESULTS OF THE ANALYSIS

When the job is completed, the user can view the results of the analysis with the Visualization module. Abaqus/CAE loads the Visualization module, opens the output database created by the job, and displays a fast plot of the model. A fast plot is a basic representation of the undeformed shape. The visualization option can be clicked in the Module list located under the toolbar; select File→ Open and select Pushover-1.odb from the list of available output database files and click OK.

To obtain the displacement outputs of the model, From the main menu bar, select Result→History output. The History Output dialog box appears.

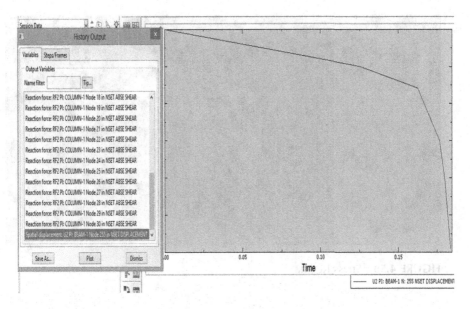

**FIGURE 4.58**   Graph of Displacement vs. Time.

Select all the base shear results at the end column nodes, and the graphs are plotted in the viewport, as shown in Figure 4.57.

Select only the displacement result, and the graph is plotted in the viewport, as shown in Figure 4.58. All the results will be extracted from this module, and further explained in the next section.

## 4.7   RESULTS AND DISCUSSION

The requirement to plot the pushover curve should be taken.

### 4.7.1   Case 1

Graph of displacement vs. time (Case1) – Figure 4.58.
   Stress result (Case1) – Figure 4.59.
   Displacement result Time (Case1) – Figure 4.60.
   Base shear force vs Time graph (Case1) – Figure 4.61.
   Resultant base shear force vs. Time (Case1) – Figure 4.62.
   Displacement vs. time (Case1) –Figure 4.63 and
   Base shear force vs. Displacement (Pushover Curve) - (Case1)- Figure 4.64

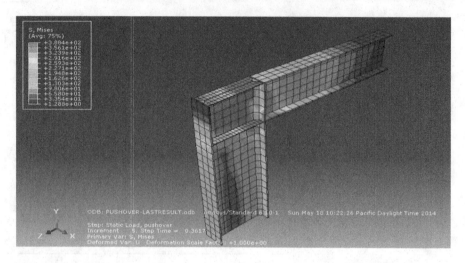

**FIGURE 4.59**    Stress Result.

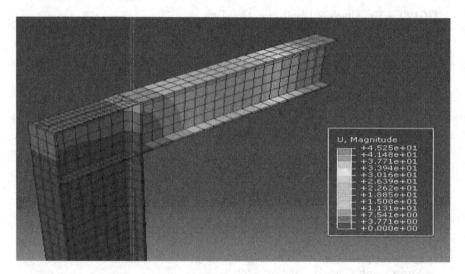

**FIGURE 4.60**    Displacement Result.

## 4.7.2  CASE 2

Base shear force vs. Displacement (Pushover Curve) (Case2) – Figure 4.65.
  Stress result (Case2) – Figure 4.66.
  Displacement result (Case2) – Figure 4.67.
  Base Shear Force vs. Time (Case2) – Figure 4.68.

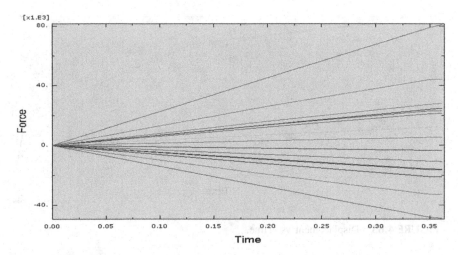

**FIGURE 4.61** Base shear force graph.

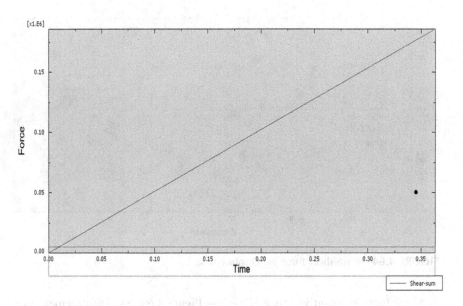

**FIGURE 4.62** Resultant Base Shear force vs. Time.

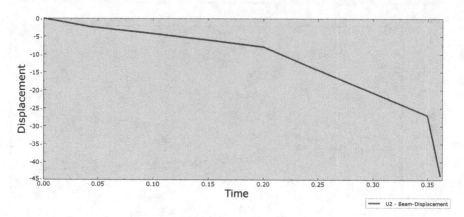

**FIGURE 4.63**   Displacement vs. Time.

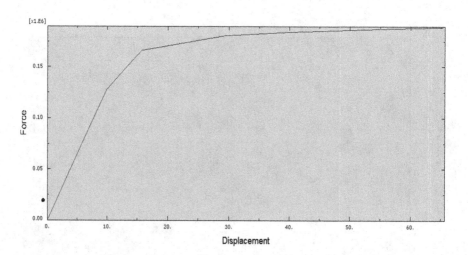

**FIGURE 4.64**   Base shear force vs. Displacement.

Shear force resultant vs. Time (Case2) – Figure 4.69. and Displacement vs. time (Case 2) - Figure 4.70

In both cases, it can be seen that the most critical area subjected to large stress is the column-beam connection. Therefore, this should be taken into consideration. It is also obvious that the peak displacement occurs at the cantilever beam.

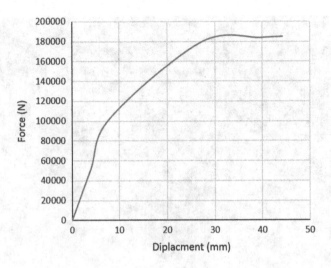

**FIGURE 4.65** Base shear force vs. Displacement (Pushover Curve).

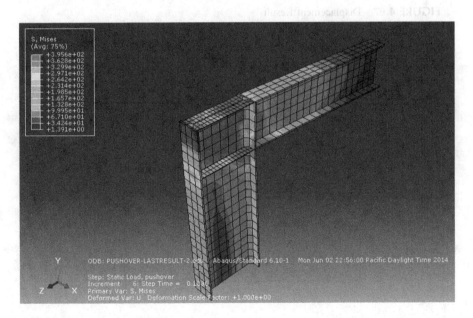

**FIGURE 4.66** Stress Result.

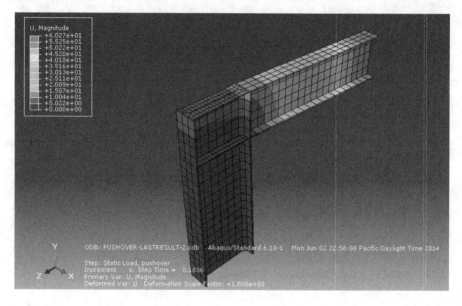

**FIGURE 4.67**   Displacement Result.

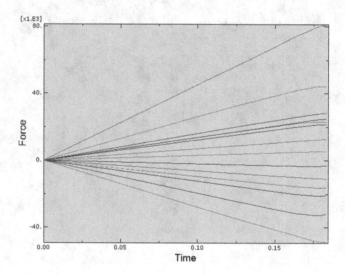

**FIGURE 4.68**   Shear force results.

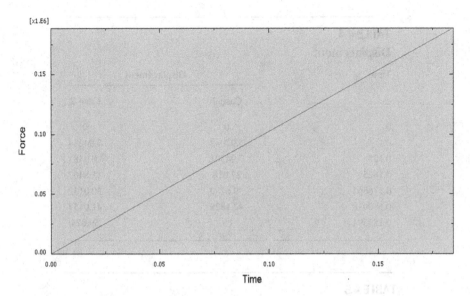

**FIGURE 4.69**   Shear force resultant vs. Time.

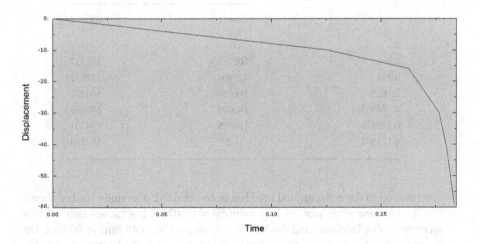

**FIGURE 4.70**   Displacement vs. Time.

## 4.8   CONCLUSION

The model has been analyzed by the pushover method to indicate the capacity of the connection between the beam and column and to find the amount of displacement and check the base stress and shear to plot the pushover curve. The results show that the distributed load is more effective over the beam, while less

**TABLE 4.4**
**Displacement**

| Time | Displacement | |
|---|---|---|
| | Case 1 | Case 2 |
| 0 | 0 | 0 |
| 0.1 | 3.97082 | 7.94284 |
| 0.125 | 7.94301 | 9.948 |
| 0.1625 | 27.018 | 15.8467 |
| 0.176563 | 39.6789 | 30.0132 |
| 0.180078 | 44.1828 | 41.0157 |
| 0.183594 | 1.1 | 59.074 |

**TABLE 4.5**
**Base Shear**

| Time | Base Shear Force | |
|---|---|---|
| | Case 1 | Case 2 |
| 0 | 0 | 0 |
| 0.1 | 102237 | 102337 |
| 0.125 | 127806 | 127932 |
| 0.1625 | 166159 | 166322 |
| 0.176563 | 180495 | 180669 |
| 0.180078 | 184029 | 184201 |
| 0.183594 | 1.2 | 187639 |

contribution to the concentrated load has been obtained. The study further shows that the increase of distributed load from 50 N to 100 N has caused only a 25 % increase in displacement and the length is increased from 45 mm to 60 mm. On the other hand, increasing the uniform distributed load, the base shear has increased by 1.4% in total base shear where it induced at the column fixed end base. Table 4.4 illustrates the displacement of both Case 1 and Case 2 at a particular time, while Table 4.5 illustrates the base shear force of both Case 1 and Case 2 at a particular time.

# 5 Strengthening of the RC Beam Using CFRP Rods and Concrete Jacketing

## 5.1 INTRODUCTION

In this chapter, the procedure for modeling of the reinforced concrete (RC) beam using carbon fiber reinforced polymer (CFRP) rods with concrete jacketing has been demonstrated. For this purpose, a conventional RC beam is considered with 150 × 250 mm dimensions, 2T12 tensile steel bars at the bottom, 2T10 bars at the top, and 8 mm links at each 200 mm. As mentioned before, the beam is strengthened by using CFRP rods and concrete jacketing at the bottom side of the beam to increase the loading capacity of the beam. So the aim is to evaluate the effect of retrofitting with CFRP on load capacity of the RC beam.

## 5.2 PROBLEM DESCRIPTION

This chapter demonstrates the various steps to model a simply supported reinforced concrete beam and also its strengthening using CFRP. In addition, we also conducted a parametric study to investigate the effect of various parameters on the behavior of the strengthened beam, as shown in Figure 5.1.

### 5.2.1 GEOMETRIC PROPERTIES

The dimensions and details of the RC beam, under strengthening, are shown in Figure 5.2. The beam before the strengthening system is a simply supported beam having a total length of 2,200 mm. The effective length will be 2,000 mm. The cross section A-A explains the reinforced details. The width of the beam is 150 mm, and the height is 250 mm. The main reinforcement will be 2T12 mm at the bottom and 2T8 at the top to hang the strips. The links are placed at each 100 mm (closed strips) with a diameter of 8 mm. The concrete cover in the horizontal direction is 15 mm and 30 mm in the vertical direction.

The strengthening system used will be CFRP rods with concrete jacketing. The CFRP rods will be fixed to the bottom of the beam using anchorage bolts. Then, 30 mm concrete jacketing will be cast at the bottom, covering the bolts and rods, as illustrated in Figure 5.3.

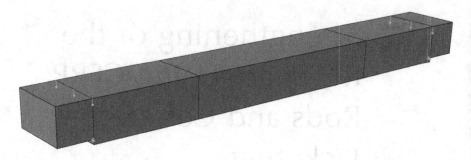

**FIGURE 5.1**   3D modeling.

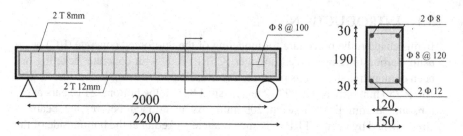

**FIGURE 5.2**   Details of the beam.

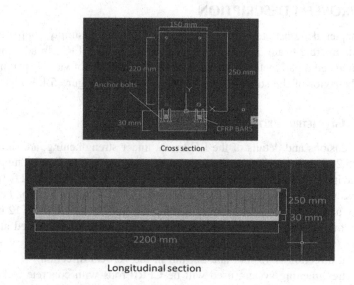

**FIGURE 5.3**   Details of the beam and strengthening system.

### 5.2.2 MATERIAL PROPERTIES

The RC concrete beam is made by concrete and steel reinforcement, and the concrete consists of cement, fine aggregate, coarse aggregate, water, in addition to some additives and admixture. The properties of the concrete rely on the properties of its initial material and its proportion in the mixture. In this modeling, concrete with compressive strength equal to 30 MPa was selected. Table 5.1 provides all the necessary information to define the concrete material.

Regarding the steel, Table 5.1 displays the properties of the steel. Two types of steel are identified; high strength steel of 500 MPa is used for the reinforcement bars and other main steel components, and for the links and secondary components 250 MPa steel will be used.

The most important material to define is the material that will be used in the strengthening system, which are CFRP rods, anchorage bolts, jacketing concrete, and epoxy as adhesive materials. All properties in Tables 5.2, 5.3, 5.4, 5.5, and 5.6 define the characteristics of these materials taken from the latest papers based on their availability.

## 5.3 OBJECTIVES

1. To create 3D finite element modeling for RC beam strengthening using CFRP rods with anchorage bolts and jacketing concrete

**TABLE 5.1**
**Concrete properties**

| Property | Value (MPa) |
|---|---|
| Compressive strength | 30 |
| Modulus of elasticity | 25,190 |
| Modulus of rupture | 308 |
| Splitting tensile strength | 2.46 |

**TABLE 5.2**
**Tensile properties of CFRP rods with fiber volumes of between 50% and 70%**

| FRP system description | Young's modulus (GPa) | Ultimate tensile strength (MPa) | Rupture strain (%) |
|---|---|---|---|
| High-strength carbon/ epoxy | 115–165 | 1,240–2,760 | 1.2–1.8 |

**TABLE 5.3**

**Steel properties**

| Reinforcement type | Yield strength (MPa) | Modulus of elasticity (GPa) |
|---|---|---|
| Reinforcement bars: Main components | 500 | 200 |
| Links: Secondary components | 250 | 185 |

**TABLE 5.4**

**Epoxy adhesive properties**

| Property | Value |
|---|---|
| Modulus of elasticity (GPa) | 3 |
| Elongation at ultimate (%) | 2.6 |
| Tensile strength (MPa) | 55 |

**TABLE 5.5**

**Jacketing concrete properties**

| Property | Value (MPa) |
|---|---|
| Compressive strength | 30 |
| Modulus of elasticity | 30,150 |
| Modulus of rupture | 4.11 |
| Splitting tensile strength | 3.28 |

**TABLE 5.6**

**Anchor bolt mechanical properties**

| Grade | Size | Tensile (MPa) | Min Yield (MPa) | Elongation (%) | Min RA (%) |
|---|---|---|---|---|---|
| 55 | 2.75–3 | 517–655 | 380 | 21 | 20 |

2. To measure how much strengthening will develop the behavior of the beam
3. To investigate the effects of different parameters on the behavior of the strengthening beam

## 5.4 MODELING

Creating 3D finite element modeling in almost all finite element programs involves similar procedures. Abaqus, like other programs, consists of a series of steps. These steps are summarized into three phases, as shown in Figure 5.4.

Figure 5.5 shows the steps involved in modeling of the RC beam strengthened with CFRP rods and concrete jacketing using Abaqus. Each step depicted in the diagram will be further clarified.

### 5.4.1 Part Module

Any real structure can be converted into a 3D model. This structure is initially divided into a number of elements or parts. The assembly of these parts will produce the structures. The modeling consists of five parts. Three parts formulate the RC beam, and two parts produce the strengthening system. The first three parts consist of the body of the RC beam, rebar reinforcement, and the stripes. The other two parts comprise the CFRP rods and concrete jacketing. The

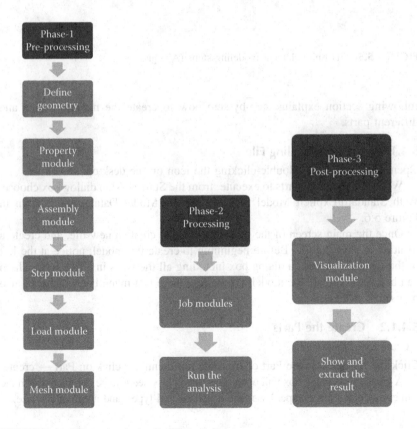

**FIGURE 5.4**   The main phases in 3D modeling.

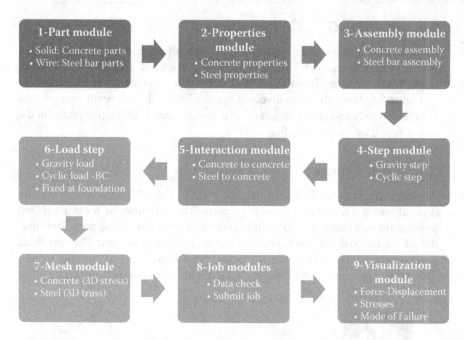

**FIGURE 5.5**   3D finite element modeling steps (Abaqus).

following section explains step-by-step how to create the modeling files and different parts.

### 5.4.1.1   Create Modeling File

Open Abaqus/CAE by double-clicking the icon on the desktop.

When the program starts to execute, from the Start Session dialog box choose, With Standard/Explicit Model from the Create Model Database as shown in Figure 5.6.

Once the main screen of the program appears, create a new file, and create a location to save the file. Before beginning to create the model, notice at the left of the window, there is a dialog box including all the steps in a tree form which can be used to create the model. At the top, there is a menu bar which can also be used.

### 5.4.1.2   Create the Parts

*i Concrete Beam*

Click the lift trees Create Part or from the top menu bar click on Part → create.

A Create Part dialog box will open. This dialog box needs to be completed with the name, modeling space, type, base feature (shape and type), and approximate size.

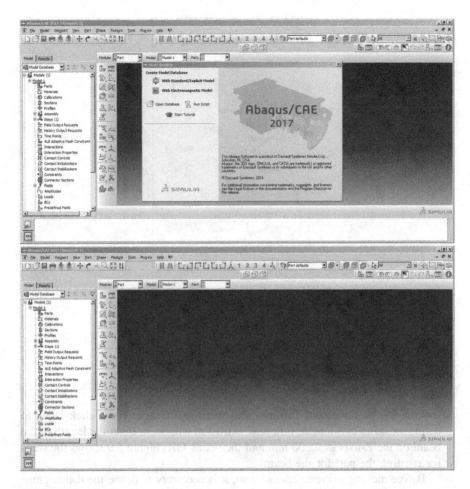

**FIGURE 5.6**   Create the modeling file.

Complete the dialog box using the information given here. Name: R-C-Beam, Modeling Space: 3D, Shape: Solid, Type: Extrusion, and Approximate Size: 10,000, as shown in Figure 5.7. Once completed click Continue to proceed to the next step.

A new space will open that will allow defining or drawing the shape of the cross section.

Start to sketch the section by solid extrusion with dimensions for the beam's cross section.

In this space, the program provides numerous drawing tools, as shown in Figure 5.8, to sketch and modify the drawing.

By using Create Lines: Rectangular (four lines), draw a rectangular shape.

Take X, Y plan for the cross section, (0,0) to Pick the starting point for the rectangular shape or enter X, Y, and (150,200) and Pick the opposite corner for the rectangular or enter X, Y. By using the Add Dimension, identify the cross

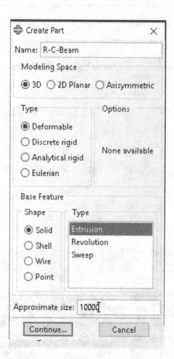

**FIGURE 5.7**   Create RC beam.

section size as horizontal 150 mm and vertical 200 mm. Then, click Done to define the other axis length. Edit base extrusion will open. Enter the length of the beam on the Z-axis as 2,200 mm and then click OK. Figure 5.9 shows the steps for creating the part for the beam.

Before moving to create another part, it is necessary to define the datum plane for the length of the beam. The purpose of this datum will be clarified later when the boundary conditions and loading are defined. Figure 5.10 shows the comparison between the R-C-Beam part in 3D before creating the datum and then after.

By using Create Datum Plane: Offset from principle plan → choose XY Plan → and make an offset of 100, 600, 1,100, 1,600, and 2,100. Then use Partition Plan: Use Datum Plane to create a partition in all previous datum planes.

Creating a partition is not necessary in the next step, and it is advised to do it later, especially in the load step. In the Load Step, as discussed later in Step load, the program will ask the user to select to Set or the surface which the loads or even the support will be assigned to.

It is recommended to create the partition while at the same time create the Set to apply the load and the support as creating the partition later will create a new part. This new part will be needed again to define its materials, and the constraints which include the beams already involved will need to be modified again during creating meshing for the beam.

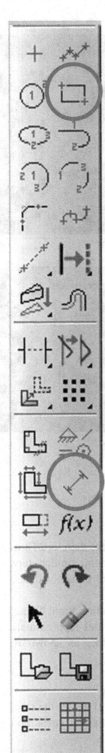

**FIGURE 5.8** Sketch tools.

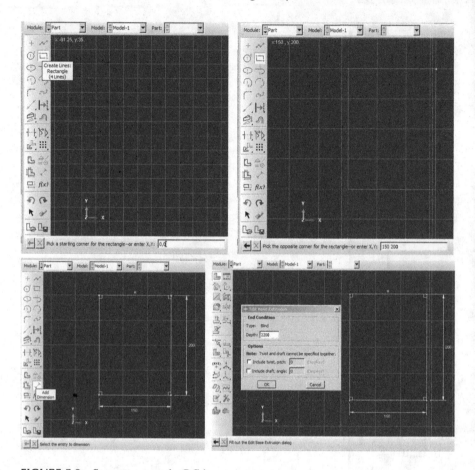

**FIGURE 5.9** Steps to create the RC beam.

## ii Bar Reinforcement

The main reinforcement is at the bottom of the 2T12. At the top, there is 2T8 to fix the link in its proper position. To create the bar reinforcement, similar steps need to be followed with minor changes that reflect the natural behavior of the steel.

The Create Part dialog box needs to include the following information: Name: Steel-bar, Modeling Space: 3D, Shape: Wire, Type: Planar, and Approximate Size: 10,000. Next, use create a line to sketch the steel-bar, as shown in Figure 5.11.

## iii Stirrups

In this beam, close T8 stirrups will be used as links as well as to gather the steel reinforcement bars. As long as the beam does not fail from shear stress and

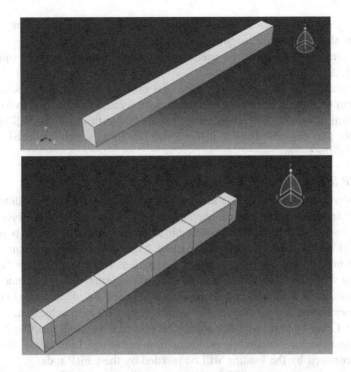

**FIGURE 5.10**  Comparison between the RC beam in 3D before and after the datum and partition.

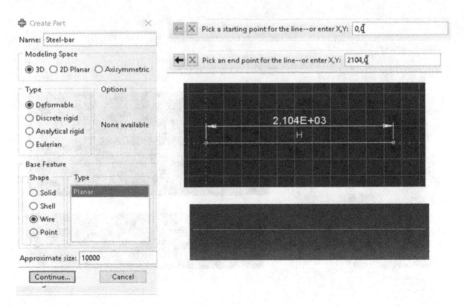

**FIGURE 5.11**  Create steel-bar steps.

according to the numerical calculation, the distance between each link will be 200 mm along the entire length of the beam.

The same approach will be used to create the steel-link part, as shown in Figure 5.12.

By creating the previous three parts, the produce or form the classical RC Beam can be done. The next parts needed to be created will be used to form the strengthening system. The proposed system applied in this example will be from the CFRP part embedded inside the concrete jacketing, which is attached to the bottom of the beam.

### iv  CFRP Rods

In this strengthened system, the main components to be used in providing extra strength and increasing the loading capacity of the beam are the CFRP rods, which provide additional tensile strength at the bottom edge of the beam. As discussed earlier, these CFRP rods will be fixed to the original beam by anchorage bolts to ensure good bonding. Later after fixing them, a new layer or jacketing will cover and protect the rod from any external effects. Another issue to be considered will be the good connection between the old and new concrete. Good preparation for the bottom side of the beam will help achieve acceptable bonding. Good bonding will also enable the system to be more functional and to avoid any immature disbanding problems from occurring. In this case, the stress to be generated by the loading will be resisted by the CFRP rods.

The steps to create the CFRP rods part are illustrated in Figure 5.13.

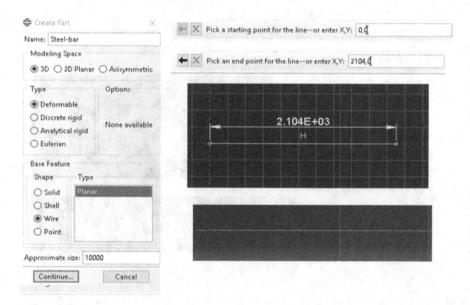

**FIGURE 5.12**  Create steel-link steps.

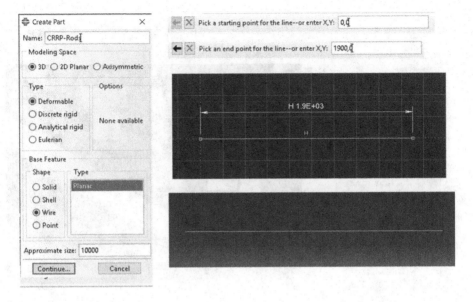

**FIGURE 5.13** Create CFRP-rods steps.

## v Concrete Jacketing

The steps to create the concrete jacketing are similar to the steps used to create the RC part. Figure 5.14 will explain the steps and dimensions that will be used.

After creating all the parts that are needed to form the modeling and before

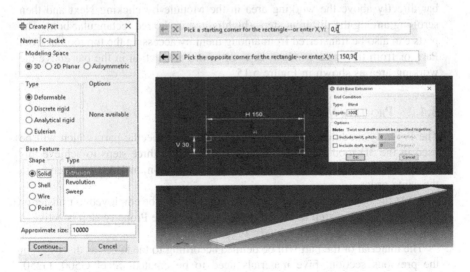

**FIGURE 5.14** Create C-jacket steps.

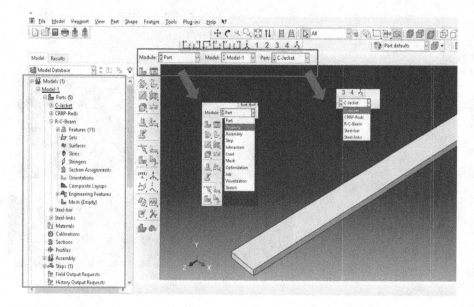

**FIGURE 5.15** Moving among the steps and parts.

moving to the second step, it is useful to learn how to transfer between the different model steps and different parts in the model.

For the model steps, as shown in Figure 5.15, there are two easy ways to transfer between steps. The first one is from the Module Database window in the left of the picture (inside the blue rectangular), and the second way is from the bar directly above the working area in the Module by clicking Next and then scrolling among the different steps (highlighted in the red rectangular box). The parts can also be transferred from among them by accessing the top toolbar from Part, or from the same window, or even from Part section in the bar above the working area as shown in Figure 5.15.

### 5.4.2 Property Module

The Property module defines the materials in each specific part, which will be used in the 3D modeling. In this program, there are three steps to achieve this target. These steps are Create Material, Create Section, and Assign Section, in the same order.

At the beginning, one of the above methods will be employed to transfer or move from the Part step to the Property step. Once the Property step is active, a window will appear to the left (see Figure 5.16).

The material of the part will be defined according to the tables as discussed in the previous section. Five materials need to be created: Steel G500, G250, Concrete G30, G40, and CFRP rods. The material will be created using the time and date of the program.

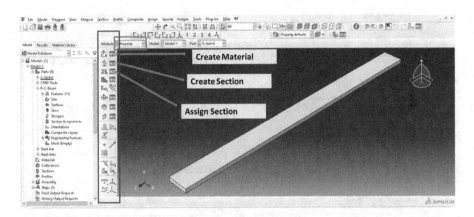

**FIGURE 5.16**   Property main steps.

### 5.4.2.1   Material Properties

*i Steel G500, G250*

As mentioned, there are three steel parts used in this model. The main reinforcement (2T12) is Steel G500 and the top reinforcement (2T8), consisting of 8 mm links is G250. In both types, linear elastic steel material is defined. According to Table 5.2 for steel G500, the modules of elasticity will be 200,000 MPa (200 GPa), but for G250, it will be 185,000 MPa (185 GPa). For the passion ratio, it will be 0.3 in both types.

In the properties window, the first option is to Create Material. Click on this option.

The Edit Material dialog box will appear. Begin to enter information into this box. Name: Steel G500.

In the material, the Editor begins to define the material parameters. Steel density, mechanical elasticity, and mechanical plasticity need to be defined by the user.

Click General → move down choose Density → in the Data, Mass Density type 7.85E-009.

Click Mechanical → Elasticity → Elastic in the material editor bar. In the data section, enter Young's Modulus as 200,000 MPa, and Poisson's Ratio 0.3.

To define the elasticity, again Click Mechanical → Plasticity → Plastic from the material editor bar. In the data section create two rows and enter the Yield Stress column as 400 MPa and 500 MPa up to dawn, and in the Plastic Strain column enter 0, 0.25 (Figure 5.17).

Finally, click OK to finish editing the material box.

For Steel G250: Repeat the same steps, except there will be a minor change to some values which will be entered in the data section, as shown in Figure 5.18.

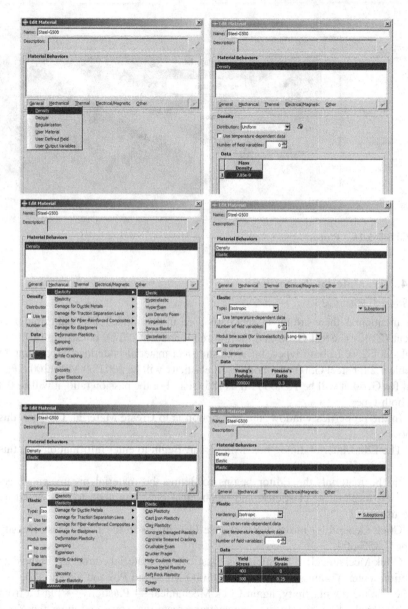

**FIGURE 5.17** Create material Steel G500 MPa.

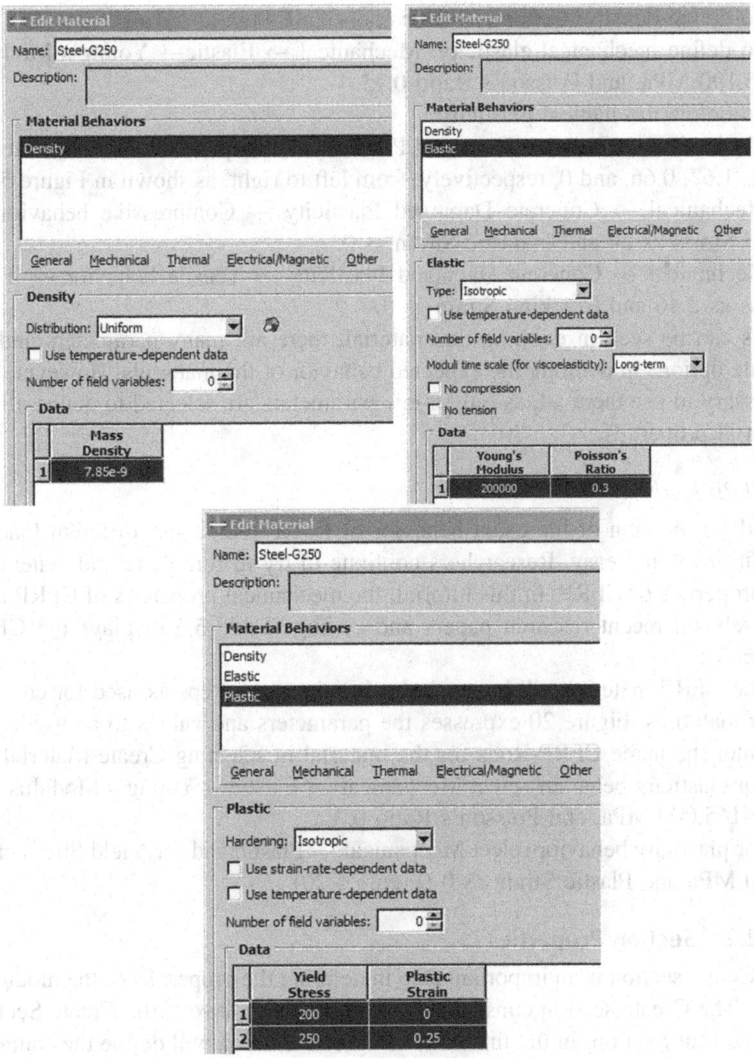

**FIGURE 5.18** Create material Steel G250 MPa.

## ii Concrete G30

Concrete G30 MPa will be used for both the concrete for the main beam prior to strengthening and in the retrofitting system as the grade for the concrete which will provide the jacketing cover. Table 5.1 identifies the essential material properties to be used in defining or creating the concrete material. The procedures to create the concrete material will be similar to the steps used in creating the steel material.

After clicking Create Material define the name as Concrete-G30.

To define density: General → Density→ 2,4E-009.

To define mechanical elasticity: Mechanical → Elastic→ Young's Modulus as 25,190 MPa, and Poisson's Ratio 0.15.

To define mechanical plasticity:

Mechanical→ Concrete Damaged Plasticity→ complete the table by entering 35, 1, 1.67, 0.66, and 0, respectively, from left to right, as shown in Figure 5.19.

Mechanical → Concrete Damaged Plasticity → Compressive behavior → Yield Stress as 24 and Inelastic Strain as 0.

Mechanical → Concrete Damaged Plasticity → Tensile behavior → Yield Stress as 2.46 and Cracking Strain as 0.

As can be seen in defining the material, there are many parameters and alternate options in defining the expected behavior of the materials. However, it is necessary to use them all, as only a few parameters are selected to define all the material properties.

### iv CFRP Rods

The determination of the exact behavior of CFRP rods under different loading conditions is not easy. Researchers continue to try to formulate and generalize the properties of CFRP. In this tutorial, the mechanical properties of CFRP rods will rely on recent research papers and coding. Table 5.3 displays the CFRP properties.

The CFRP material will be created using the same steps as used for creating other materials. Figure 20 expresses the parameters and values to be used.

Enter the name CFRP-Rods for the material in selecting Create Material.

For elasticity behavior select Mechanical → Elastic→ Young's Modulus and enter 165,000 MPa, and Poisson's Ratio 0.3.

For plasticity behavior select Mechanical → Plastic and for Yield Stress enter 1,860 MPa and Plastic Strain as 0 (Figure 5.20).

### 5.4.2.2  Section Properties

The Create section is an important step in defining the properties of the module's parts. The Create section consists of two main dialog boxes, the Create Section and the Edit Section. In the first box, the Create Section will define the Name of the section. Here, select the Category among Solid, Shell, Beam Fluid, and Other, and next determine the Type of section according to the category selected. For example, for the solid category, following types are: Homogenous, Generalized Plan Strain, Eulerian, or Composite. For the Beam category, there are two types: Beam and Truss, and so forth for all the other categories.

When completed, create the Section box and click Continue. An Edit Section dialog box will appear. The parameters in this window will be dissimilar depending on what has been chosen in the previous box. Figure 5.21 shows the two Create Section dialog boxes. In this modeling, there are six sections which are Main-Bar, Top-Bar, Link, C-Beam, CFRP Rods, and C-Jacketing.

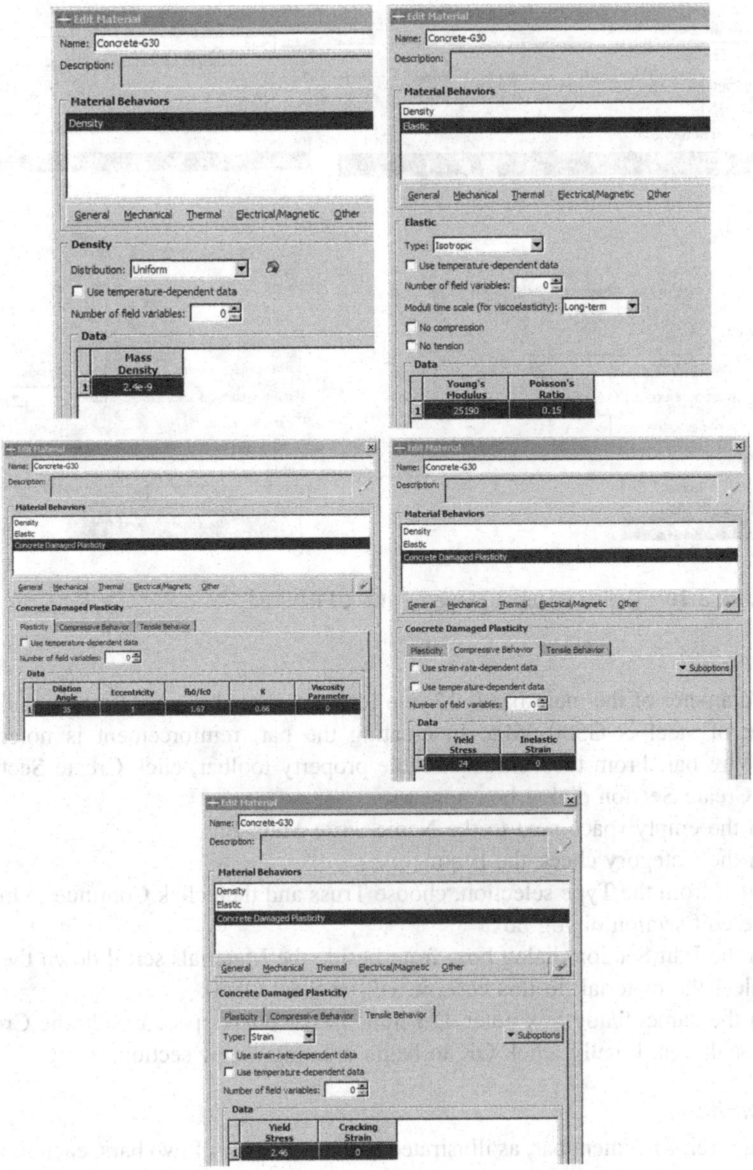

**FIGURE 5.19** Create material concrete G30 MPa.

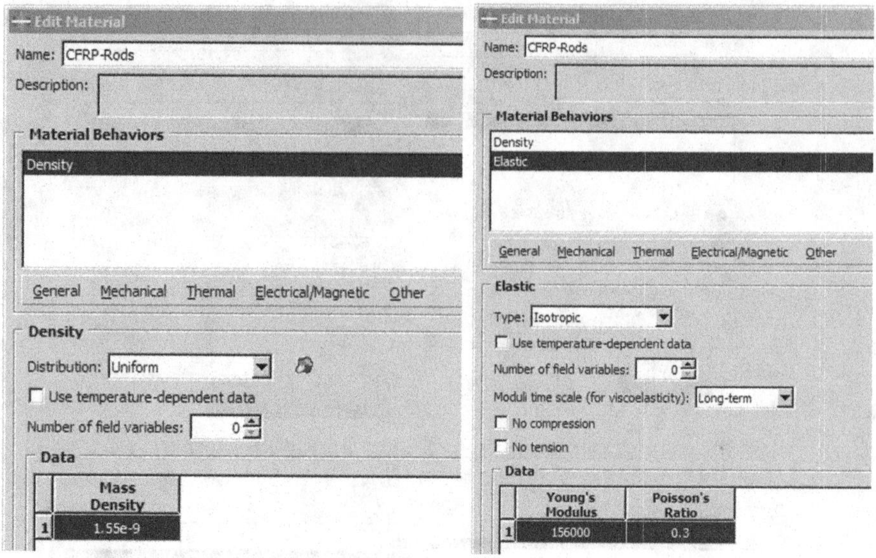

**FIGURE 5.20**   Define material properties for CFRP Rod.

*i Main-Bar*

The diameter of the main bar is 12 mm (cross-sectional area 123 mm$^2$), and the grade of steel is G500 MPa. In creating the bar, reinforcement is noted as the Wire bar. From the vertical module property toolbar, click Create Section. The Create Section dialog box appears.

In the empty space next to the Name write Main-Bar.

In the Category check the beam.

Then from the Type selection, choose Truss and then click Continue to move to the edit section dialog box.

In the Edit Section dialog box, from beside the Materials scroll down the list to select the material. In this case, it will be Steel G500.

In the same dialog box enter 123 mm$^2$ in the blank space beside the Cross-sectional area. Finally, click OK to begin creating a new section.

*ii Top-Bar*

The top reinforcement bar, as illustrated earlier, consists of two bars, each having a diameter of 8 mm (cross-sectional area 55 mm$^2$). This will be similar to the Main-Bar; (Steel G500 MPa), and in creating the Wire part.

The same steps will be applied to create the cross section for the Top-Bar except for the Cross-sectional area, which will require a space of 55 mm$^2$ instead of 123 mm$^2$.

*iii Links*

The stirrups or the links in this module are created as close link parts. The cross-sectional area will be 55 mm$^2$ and material as Steel grade G250 MPa. To define

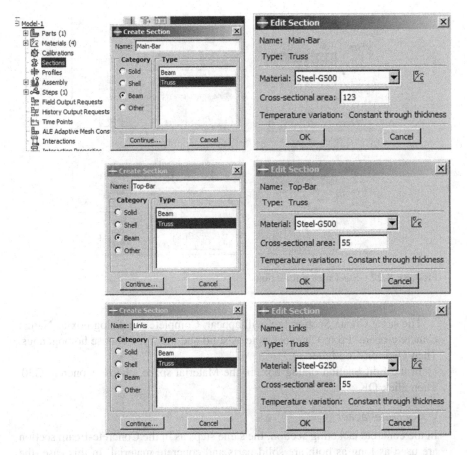

**FIGURE 5.21** Create section boxes.

the section for this link, a similar step will be followed as applied for the Main-Bar and Top-Bar with a minor change in the material as SteelG250 and the cross-sectional area as 55 mm². Figure 5.22 shows the different selections for entering the values for the steel material parts.

### iv Concrete Beam

The main section is the Concrete-beam section. After defining all the sections related to steel, the procedure to create the concrete section will begin. The Concrete beam previously created is a solid part, and the material used was concrete of grade 30 MPa. By using this data, follow the same steps to create the section but applying a different method.

From the Main Toolbar select Section, and from the sub-tree dialog box click Create.

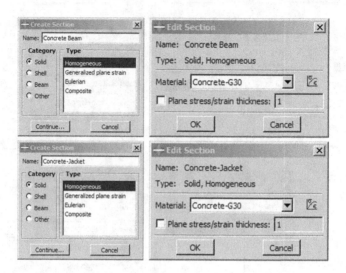

**FIGURE 5.22** Steel material sections.

The same Create Section box will appear. Complete the dialog box as Name: Concrete-beam. From Category note as solid and the Type choose homogenous. Then click on Continue.

In the Edit Section dialog box, in the Material space, choose Concrete-G30. Then click OK.

### v Concrete-Jacket

In the concrete jacketing section, the same steps as in the Concrete-Beam section are used as long as both are solid parts and concrete material. In this case, the name of the section will be Concrete-Jacket.

The Create Section and Edit Section dialog boxes for both the concrete-beam and concrete-jacket are illustrated in Figure 5.23.

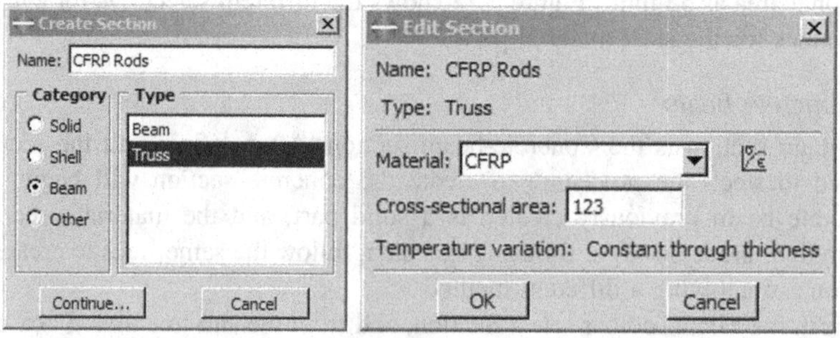

**FIGURE 5.23** Concrete material sections.

## vi CFRP Rods

The last section to be created is the CFRP rods section. The CFRP rods are created as the Wire part, and the material created as CFRP rods. In this module, the CFRP rods have a length of 1,900 mm, a diameter of 6 mm, and a cross-sectional area of 123 mm$^2$.

The same steps are applied from the Create Section dialog box and in the Edit Section box enter: Name: CFRP-Rods, Category: Solid, and Type: Truss.

In the Edit Section dialog box, choose CFRP-Rods for Material, and in the Cross-sectional area enter 123 mm$^2$ (see Figure 5.24).

### 5.4.2.3   Assign Section

The Assign Section is the third and final step in defining the modulus properties. In this step, after creating different materials as well as the required sections, allocate the creating sections to the different modulus regions. Assign each defined section to its corresponding part(s) or as a partial region of these parts. Abaqus provides an option to create a set for the region of all or partial region of the parts. To assign the section for different parts, the following steps are applied:

In starting the assign procedure, there are many ways to start the Assign Section. The prompt area will also help the user.

From the main Menu select Assign and then click Section.

From the property vertical toolbar choose Assign Section.

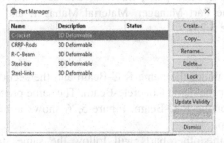

**FIGURE 5.24**  CFRP-rods material section.

From the left window, select Part to select the specific part. To expand the Menu → select Section Assignment.

The prompt area will ask the user to Select the Region to be Assigned a Section and provide an option to Create A New Set for this region and request a Name.

By using the mouse pointer, select all parts or a certain region of the parts. Select the part or the region either by clicking on the region or creating a window around the part using the mouse.

After selecting the part or all the regions click on Done. A new dialog box will appear under the name Edit Section Assignment. In the prompt area, it will ask the user to complete the dialog box.

The edit section assignment will create a section to be assigned for the selected region. This list appears when clicking on the small arrow next to the Section. Then, from this list, select the section which corresponds to each region.

Once clicking on the selected section, in the same box, Material and Type will appear. Check them both and click Ok to confirm selected section.

The dialog box will then disappear, and the selected region will convert to a green color, where the prompt will then ask the user to start perform another assign for a new region.

Repeat the same steps for all parts and the region required in the 3D modeling,

Figure 5.25 displays a summary for all created Parts, Materials, and all created Sections. In addition, Abaqus provides the ability to modify, copy, rename, and even delete all created parts, material, and sections. This can be easily performed using the Part Manager, Material Manager, and Section Manager, respectively (Figure 5.25).

*RC-Beam*

The part is created under the name R-C-Beam and the section assigned to this part is created under the name Concrete-Beam. The same procedures are applied to Create Assign for the R-C-Beam. Figure 5.26 shows a number from 1 to 5 explaining each step.

For assign, all the other parts will follow the same steps, as shown in Figure 5.27.

### 5.4.3 ASSEMBLY MODULE

Assembly in the model means collecting all the parts which have already been created and were previously defined as material having a certain configuration in establishing the 3D model.

In this tutorial, as previously mentioned, it will present the ordinary reinforcement beam and strengthening system. Three parts are involved in creating

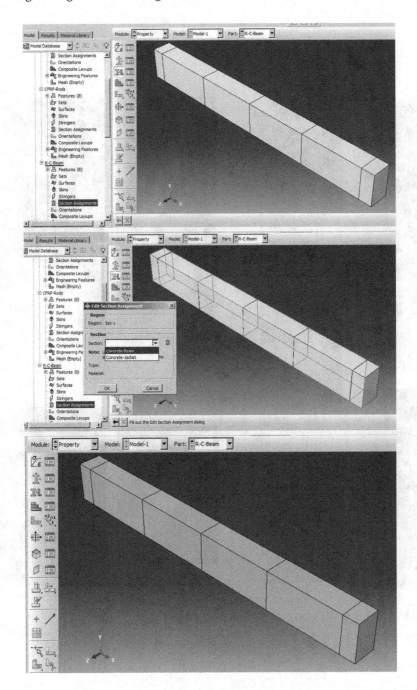

**FIGURE 5.25** Part, material, and section manager.

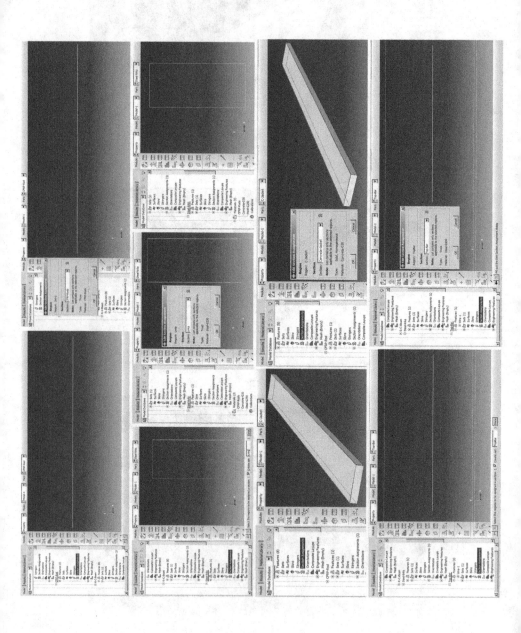

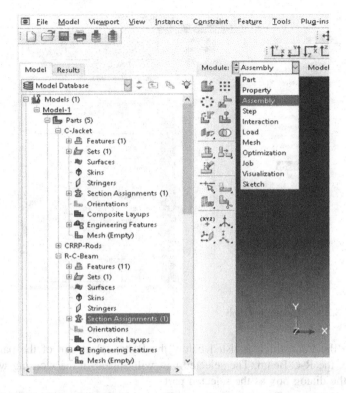

**FIGURE 5.27** Assign for all parts.

the ordinary RC beam: R-C-Beam, Steel-bar, and Links. Whereas, only two parts are involved in the strengthened system: CFRP-Rods and C-Jacketing.

### 5.4.3.1 Assemble Part Instances into the Modeling

In the module toolbar, above the working area, change the list to Module Assembly. Similar to other steps, the vertical toolbar and the main toolbar at the top will immediately change, as shown in Figure 5.28.

In the assembly toolbar, there are numerous supporting tools. These tools will help create accurate modeling like Linear Pattern, Translate Instance, Move Instance, Edit Feature, and so forth. To start from the module track, the following steps are performed:

Insert all the parts or instances that are required to build the module. This is achieved in several ways, as shown in Figure 5.29.

From the main top toolbar click on Instance, and then select Create. Or from the vertical toolbar, click the first icon, which is Create Instance.

A Create Instance dialog box will appear. The displayed prompt will request the user to Select the parts/modules to Instance from the dialog box.

From the dialog box, choose the parts below the Create instances.

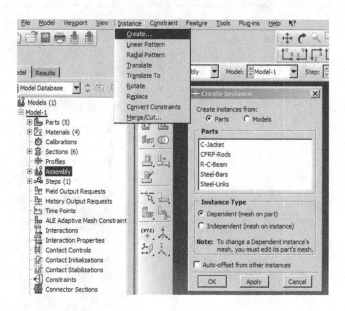

**FIGURE 5.28**  Module assembly.

From the Part which is listed below the Parts, select one of the parts, for example, the R-C-Beam. The selected part will appear in the working window behind the dialog box as the selected part.

In the Instances Type option choose Dependent (mesh on the part).

Then click OK to finish or Apply to move to create an instance for another part.

Be sure to create an instance for all parts, as shown in Figure 5.30.

Start to use the supporting tool in the vertical assembly toolbar to move and locate each part in its proper position and direction.

Starting with the R-C-beam, its position and direction are ok. This includes the right bottom edge in the coordinator (0,0,0) and its cross section in the XY plane, and the longitudinal section on the Z axis.

For the second part, C-Jacketing, its direction is ok, but the location needs to be moved to the position between the support directly under the bottom surface of the C-R-Beam.

To perform this step, use translate instance tools → select C-Jacket when asked and Select instance to translate. Then select from the top right edge when asked to Select start point for translation vector—or enter (X, Y, Z) → by entering (0,0,100). When asked to Select the end point for translation vector—or enter (X, Y, Z) → click on ok when asked about the Position of instance if it is ok to translate or type X to delete the translate and move back if there is a need to modify or change. When it is ok, the C-Jacket will move to its accurate position and appear in the new position. Figure 5.31 shows the treatment for the C-Jacket.

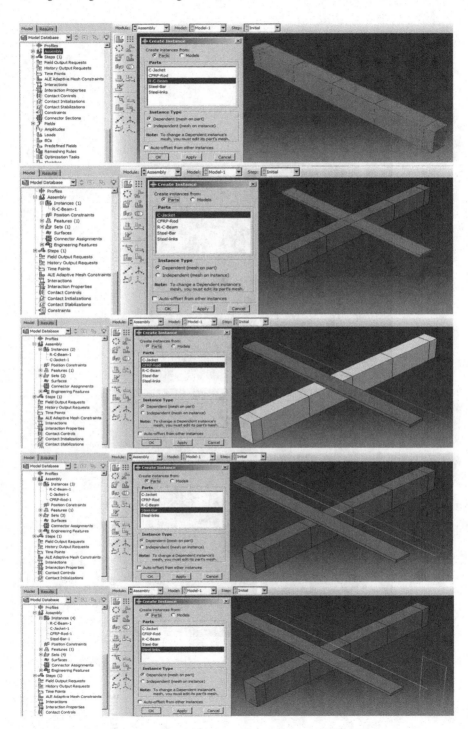

**FIGURE 5.29** Create instances.

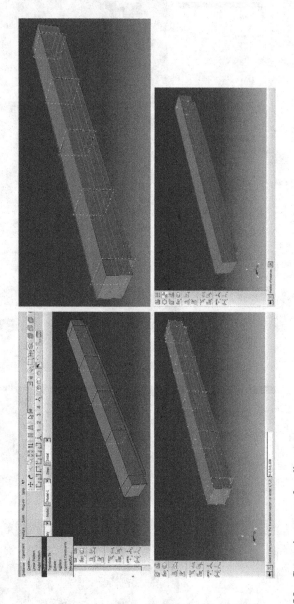

**FIGURE 5.30**  Create instances for all parts.

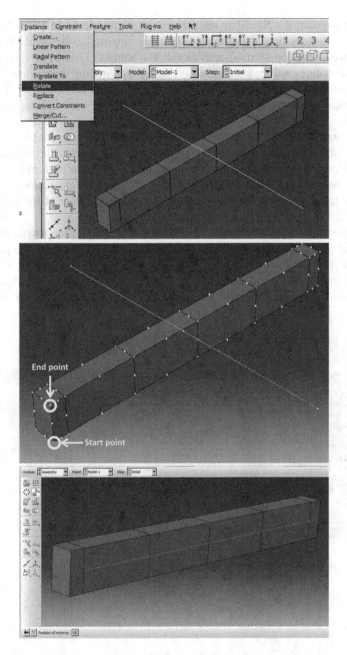

**FIGURE 5.31**  Treatment of jacketing.

At this point, the Steel-bar instance is not correct, even in the location or direction. Besides that, the number of bars is supposed to be depicted as four bars. Two at the top (8 mm), and two at the bottom (12 mm) at each corner. Many procedures will be used to resolve this issue. Starting from rotate instance and then translate instance and finally, linear pattern. All these tools are found in the vertical assembly support bar or can extend from the main top bar from Instance.

To Rotate instance: Select the steel-bar instance when prompted by Selecting the instances to rotate after clicking on the rotated instance. After selecting the Steel-bar, click Done. The same conversation will begin in performing the operation. When asked about Select a start point for the axis of rotation—enter X, Y, Z by inserting (0.0,0.0,0.0) → choose the right top corner of the R-C beam (Figure 5.32).

Or alternatively, enter (0.0,250.0,0.0) as the axis for rotation when prompted to Select an end point for the axis of rotation—or X, Y, Z. Then the program will ask the user for the angle of rotation and show the new position. In this case the angle will be -90° for rotation. Enter (-90°) and click enter. Again, the instance will appear in its new position. Click OK when asked if the Instance

**FIGURE 5.32** Rotated steel-bar.

Position is correct or X to return to make any changes. Figure 5.32 displays this rotation.

Translate instance: Moving the steel bar from its new position after completing the rotation process is finished is the second procedure for the steel bar. In Figure 5.31, the steel-bar in its new position is almost inside the body of the R-C-Beam and cannot be seen or selected easily. So, before moving to translate the steel bar, some treatment is needed. From the main Menu go to View extend it and select and click on Assembly Display Option. The Assembly Display Options dialog box will appear. Click on Instance from the various tabbed pages on the top of this box. From here, the user can control which instance is visible in the window and which one is hidden by clicking in the Visible Box beside each instance. In this case, as illustrated in Figure 5.33, R-C-Beam-1 and C-Jacket will be hidden. Then click Apply to show the change or click OK to close the dialog box.

The steel-bar becomes visible using the same procedure. The steps for Translate C-Jacket will be repeated here except when asked about Select end point for translation vector—or enter (X, Y, Z) in which it will be (15.0,30.0,30.0), as shown in Figure 5.34.

Linear Pattern: After checking the steel-bar in the correct position and direction, next create the linear pattern for the steel reinforcement. As long as there are four bars, and the distance between them horizontally is 120 mm and vertically 190 mm the same distance will be used. From the vertical assembly supporting bar, the linear pattern will be chosen by clicking on it. Then click on the Steel-bar when the prompt appears asking to Select the instance to pattern and then click Done. A linear pattern dialog box will appear and ask the user to determine the Number of Offsets on Direction1 and Direction2 and the distance between both. As mentioned early, the number of patterns will be two in both directions and 120 mm and 190 mm in the horizontal and vertical directions, respectively. The new creating patterns dialog box will appear in the window

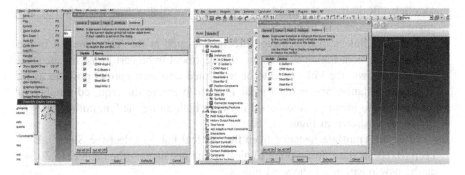

**FIGURE 5.33** Assembly display option.

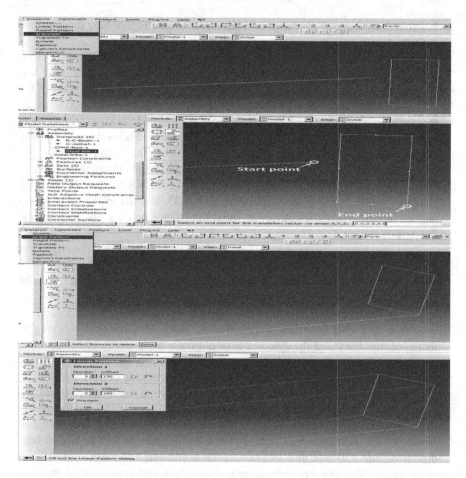

**FIGURE 5.34** Translated steel bar.

→ click OK if it is acceptable or return to the dialog box to carry out any changes, as shown in Figure 5.35.

The steel link is in the correct direction located at the XY plane but in the incorrect position. Figure 5.36 shows the treatment to be performed for the steel links. As shown, the links need to be translated into their correct position using Translate Instance, and using Linear Pattern to create the required number of links. Also, to identify the distance between both along the longitudinal axis of the beam as shown in Figure 5.36.

Click on Translate Instance from the vertical bar → select the Steel-link → keep (0,0,0) as the starting point to pick up the instance → then enter (15,30,30) as the new position for the steel link.

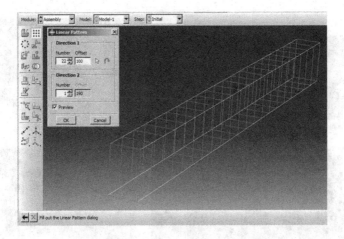

**FIGURE 5.35**   Create a linear pattern for the steel-bar.

Click on Linear Pattern from the vertical bar → select the Steel-link → fill the box with 22 as the number of offsets with a distance equal to 100 mm to change the direction of offset along the Z axis.

The CFRP bar is the last instance in need of treatment to place it in its correct direction and position. Similar to the steel bar, the CFRP bar will need to go through three processes, namely rotated instance, translate instance, and, finally, linear pattern.

Click on rotate instance → select the CFRP bar to identify the two points in the vertical edge of the R-C-Beam as the axis of rotate → enter (-90°) as the angle of rotation.

Click on translate instance → select the CFRP bar → keep (0,0,0) as the starting point for transfer and enter (15,-15,180) as the new point for the CFRP.

Click on linear pattern → select CFRP bar → complete the dialog box by entering 2 as the number of offsets with 120 mm in the X direction (Figure 5.37).

After assembling and gathering all the instance in their correct direction and position, the finite element is ready and almost reflects the real structure. Abaqus provides a variety of views for modeling and for different cutting sections. Furthermore, many alternate options for coloring codes are available, as shown in Figure 5.38. All these options can be found on the main menu bars above the working area.

### 5.4.4   STEP MODULE

The Step module is considered as one of the essential steps in 3D simulation modeling. The main parameters which are needed are found in the step toolbar, namely, Create Step, Create Field Output, and History Output. In each, there are many parameters which need to be determined. These parameters are different for every model, depending on the types of analysis required, and results or

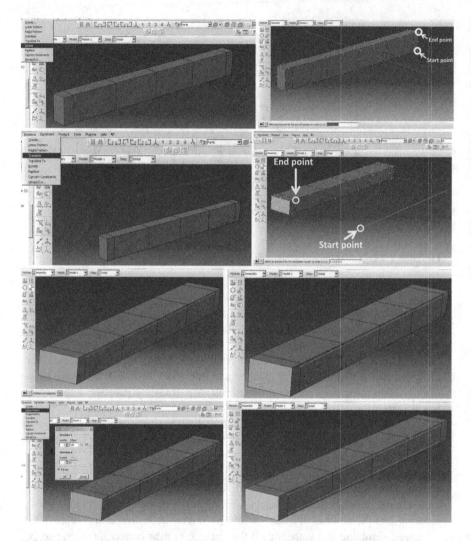

**FIGURE 5.36** Translate and linear pattern for the steel-link.

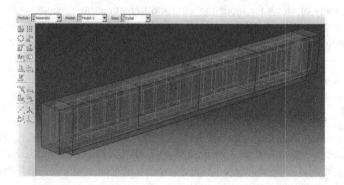

**FIGURE 5.37** Rotate, translate, and linear pattern for the CFRP bar.

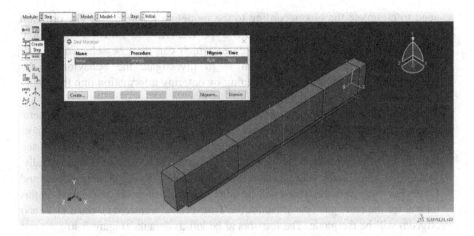

**FIGURE 5.38**  Different color codes for 3D assembly modeling.

output required. To begin, from the top bar above the working area, scroll to Module and then select Step.

### 5.4.4.1  Create Step

The program automatically creates the first step in the analysis under the name Initial Step and the procedure (Initial) can be used by clicking on the step manager dialog box, as shown in Figure 5.39. To create another step, the user can click on Create Step → then the Create Step dialog box will appear. The user will

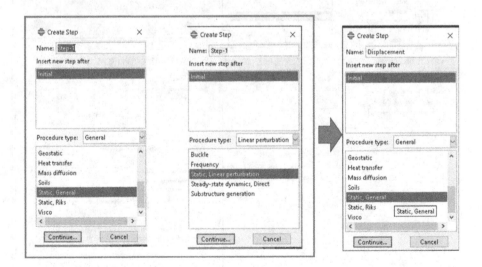

**FIGURE 5.39**  Initial step.

need to complete the box according to the modeling requirement. The program with this addition provides Procedure Type General and Linear Perturbation. Each will provide list of loading cases which can then be selected through scrolling. In this module, the aim is to investigate the behavior of the strengthening system under static loading by entering information into the dialog box, as shown in Figure 5.39.

In the Name enter Displacement instead of the default Step-1. Select General from the Procedure Type → and select Static, General → and then click Continue.

The Edit Step dialog box will appear after clicking on continues. In the edit step box, the name Displacement and type Static, General will be generated automatically. In addition, three gates will appear in this box, namely Basic, Incrementation, and Other. For each gate, the values will be different from one model to the next. In this tutorial for the basic gate, the value for the Time period is entered as equal to 10 and NIgeom will be the mode. The rest should be left as defaults (Figure 5.40).

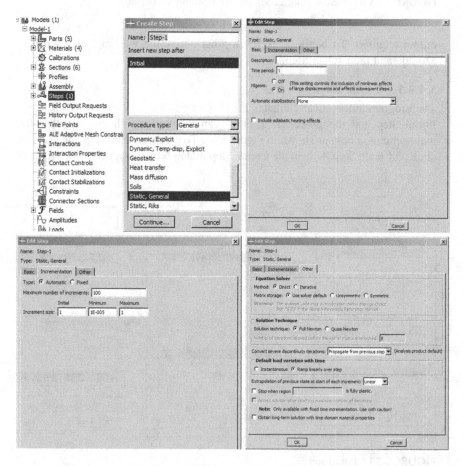

**FIGURE 5.40**  Creating step.

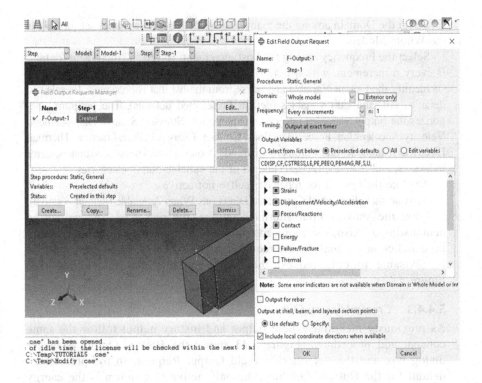

FIGURE 5.41   Edit step.

The Incrementation gate is the most significant setting in this step, which controls the analysis being performed. The incrementation gate will request the user to identify the Maximum Number of Increments and Initial, Minimum, and Maximum increment size. In the beginning, the user can enter a value and later can change the values in case there are any problems in the analysis. For more advanced settings in the step, Other gates can be used, although in most cases it should be left without any change. The three gates and their values are shown in Figure 5.41.

### 5.4.4.2   Create Field Output

The Create Field Output or Field Output Manager will manage and control the results or outputs which are expected from the analysis. Abaqus provides an option to show the results and select exactly what results are required, suggesting the points to obtain the results.

The program will create F-Output-1 for the Displacement step. When clicking Edit in the Field-Output Request Manager dialog box, the program will ask the user to identify the following:

Select the Domain among the many choices as the whole model set (default is the Whole Model).

Select the Frequency among the Last increment: Every "n" increment (default is every n increment, with n = 1).

Identify the Output Variable and select from the list that is presented: Preselected defaults, All, and Edit variables (default Preselected defaults). The program categorizes the output variables into certain categories: Stresses, Strains, Displacement/ Velocity/Acceleration, Forces/Reactions, Contact, Energy, Failure/Fracture, Thermal, Electrical /Magnetic, and Porous media/fluids, Volume/Thickness/Coordinates, Error indicators, and State/Field/User/Time all of the variable output categories.

Activate the Output for rebar (default = not active).

Activate Including Local Coordinate Directions When Available (default = active).

From the above output alternate options, the power of Abaqus can be demonstrated. Also, the more complicated the requested output, the more time and complex analysis are required, and vice versa. From view point, it is advisable to only request the output variable that is required (see Figure 5.42).

### 5.4.4.3   Create History Output

As previously mentioned, field output and history output follow the same concept and procedures. Figure 5.43 shows the History Output Manager dialog box and the default Edit Field Output Request. In the preselected default for the Output Variables, the only active box shown is the energy indicator. This is because the History Output is more concerned about the user selecting the results more than the default analysis results. After finishing all simulation requirement steps, the user returns here to confirm the required result.

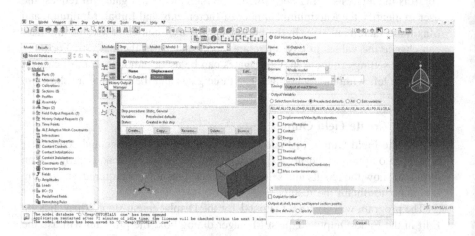

**FIGURE 5.42**   Edit field output request.

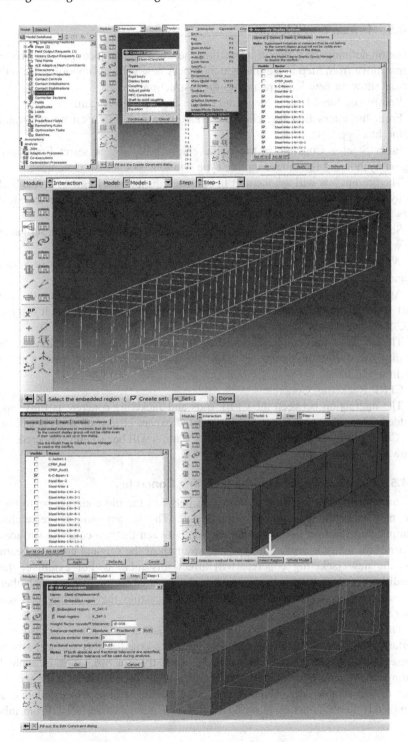

**FIGURE 5.43**  Edit history output request.

## 5.4.5 Interaction Module

To begin defining the different interactions between the instance from the Model Steps, move from Step to the Interaction step. A new vertical supporting bar will appear showing new icons related to the instruction.

Reinforced Steel Bars + Steel Links with the Concrete Beam:

From the interaction bar, click on Create Constraint → a create constraint dialog box appears.

Enter Steel + Concrete in the Name empty box → then select Embedded region by holding the mouse on the Type window → click on Continue.

The prompt box will ask the user to Select the Embedded Region (Create Set) → by using Assembly Display Option close all the instances except for the steel bars and links, as shown in Figure 5.44.

Select the steel reinforcement and type Steel Reinforcement as the set name → when finished click Done.

When asked about the Selection Method for the Host Region choose Select Region.

Again, in the prompt, it will ask the user to Select the Host Region (Create Set) → which now displays the R-C-beam only in the window. Select it → type R-C-Beam as the set name → click Done.

The Edit Constraint dialog box will appear to review the previous choices. To modify the constraint parameter go to the Default section and click OK to confirm and finish.

CFRP Racketing Concrete

The interaction between the CFRP rods and Jacketing concrete is fully embedded. Thus, the same procedure is repeated except instead of steel, it will be the CFRP rods, and instead of the concrete beam, it will be jacketing concrete, as shown in Figure 5.45.

### 5.4.5.1 Concrete Beam with the Jacketing Concrete

As mentioned, the relation or interaction between the old concrete beam and the new jacketing concrete is complicated. The reasons which create this complicated reaction is that the interaction between the two concrete surfaces depends on many factors. The type and strength of the original concrete beam and the new jacketing concrete will affect this relation. For example, the conditions of the bottom concrete surface due to presence of voids and whether the surrounding beam is even at the time of casting the jacketing or even after that.

The surface between the original and new jacketing concrete is also important. The debonding phenomenon is mostly accrued at this surface, which causes an effective strengthening system. Industry practice, as it is applied during strengthening the system, is when they begin pretreatment on the tensile side of the reinforced concrete beam. This treatment aims to ensure that a completely tied interaction is produced between these two surfaces. From this

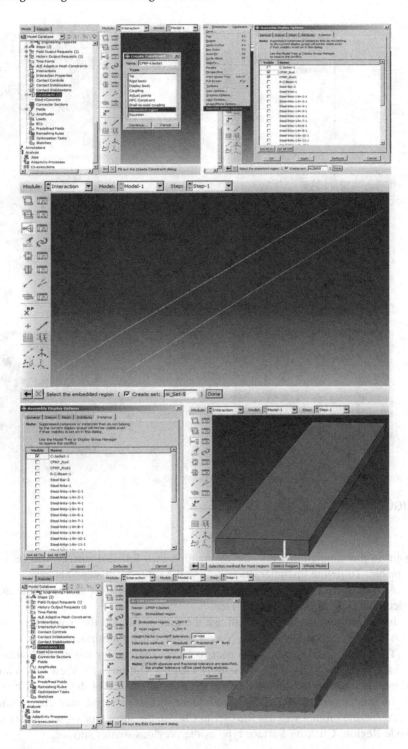

**FIGURE 5.44** Create steel concrete constraint.

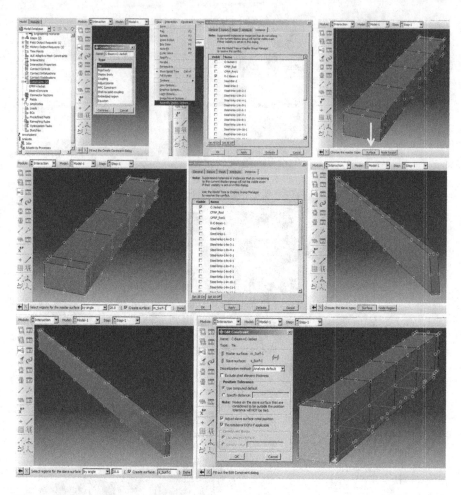

**FIGURE 5.45**  Create CFRP jacket constraint.

point in the simulation work, the assumption of a full tied surface will be acceptable. To define this interaction, follow the subsequent steps below:

Again, create a new constraint with C-Beam + C-Jacket as the Name and select the Type of interaction as Tied.

In the prompt, the program will ask the user to Choose the master type and provide two options Surface and Node Region. Click on Surface.

The prompt will then ask the user to Select regions for the master surface (create surface...). By using Assembly Display Options and View and View Manipulation Bar, the user can select the bottom surface of the original beam as the master surface and give it (the bottom surface) a name, then click Done.

Again, the prompt will ask the user to Choose the slave type either Surface or Node Region. Click on Surface type as the previous master surface.

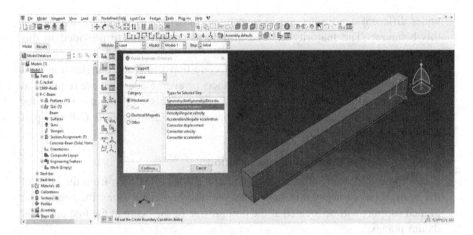

**FIGURE 5.46** Create C-beam and C-jacket constraint.

Use the same tools Assembly Display Options and View and View Manipulation Bar to select the top surface of the jacketing concrete as the slave surface when the program asks the user to Select regions for the slave surface (create surface…). Here, in the Create Surface empty box open Top Surface and then click Done.

The Edit Constraint dialog box will appear in the window. Review the Name of the constraint, its Type, and the selected surface as the Master and Slave surface. This box also provides the option to select from among the Discretization method and Position Tolerance. Review and modify or keep as the default and click OK to confirm creating the constraint.

At any time, the user can review and edit or even create a new constraint by clicking on the Constraint Manager icon and from Constraint Manager dialog box, as shown in Figure 5.46.

### 5.4.6 LOAD MODULE

Up until the previous steps, all concerns were with regard to defining and identifying the general section geometry and its materials and how to define the interaction between the components of the modeling. In this step, the main purpose of the modeling is accomplished in this section. What exact output result is needed for each loading configuration, the condition of the assembly, and its relationship with other external members will be defined.

This step is divided into two main sub-steps. The first sub-step is identified as the boundary condition of the modeling. The boundary condition reflects the position, direction, and type of surrounding support. Abaqus provides a vast range for the different types of boundary conditions to represent the real situation. The value of the boundary condition follows the same unit system as defined at the beginning. The second sub-step is defined as the load configuration. Different kinds and type of load are applied to study the behavior of the system under different conditions and to investigate the effect of different variables in this behavior. The majority of load kinds and conditions are provided in this program.

Before starting to assign the boundary condition and load configuration, first locate their position by a certain Set, Surface, or even Nodes that are required to be described and to extract the required results. This explains the reason for creating many datum planes in a different distance during the R-C-Beam part creating process.

### 5.4.6.1 Define the Boundary Condition

From Figure 5.2, each support is located at a distance 100 mm from the edge of the beam. Five datum planes have been previously created along the beam. The pinned support will be located at the bottom side of the beam at the first and final datum planes.

To assign the pinned support and their place, follow the steps below as described below.

Transfer or move from the interaction step into the Load Step from the scrolling list in the model steps. When the load model shows the vertical toolbar, a number of icons will appear.

From the top bar or vertical toolbar, click on the Create Boundary Condition icon → a Create Boundary Condition dialog box will appear. In the prompt, it will ask the user to complete the box by requesting the name, step, and type of boundary, as shown in Figure 5.47.

For the Name, enter *Support*.

In the Step, by scrolling in the narrow, two steps Initial and Displacement are

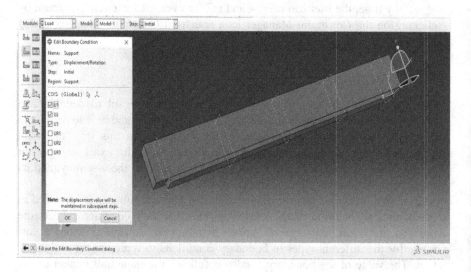

**FIGURE 5.47** Create boundary condition.

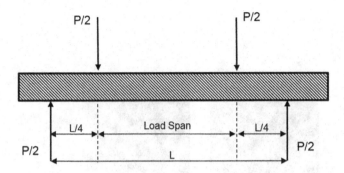

**FIGURE 5.48** Assign and edit boundary condition.

seen, which are already created in the Step Model stage. The boundary starts from the initial step; keep it as Initial.

The program (Abaqus) will provide different boundary condition Categories. The program divides these into four categories, namely Mechanical, Fluid, Electrical and Magnetic, and Other. Here, select Mechanical.

After selecting Mechanical from the boundary category, several types will be shown. Under Types for the Selected Step, many types will be listed. From this list, choose Displacement/Rotation → and then click Continue.

The program will then ask the user to determine the region to assign the boundary condition. The prompt will ask to Select regions for the boundary condition (Create set…). Use the Views Manipulation bar to select the bottom edge of the first and the final datum planes, as shown in Figure 5.48. Then Create a set for them under the name Support and then click Done.

The Edit Boundary Condition Dialog box will appear. From this box, review the Name, Type, Step, and Region. The control of freedom and release of rotation and displacement of the support can be modified from this dialog box. Where U1, U2, and U3 are displacements in the X, Y, and Z global axis, respectively, and UR1, UR2, and UR3 are the rotations in X, Y, and Z, respectively. For pinned support, there is no constraint for the rotation. Therefore, click on the displacement as shown and then click OK.

### 5.4.6.2 Define the Loading Configuration

The loading configuration is in accordance with the ASTM standard, ASTM D6272-02 (Standard Test Method for Flexural Properties of Unreinforced and Reinforced Plastics and Electrical Insulating Materials by Four-Point Bending). As shown in Figure 5.49, one half of the support span will be used to conduct the tests. The clear distance between the support will be divided into quarters where each quarter will be equal to 500 mm. The loading is according to the ASTM standard and applied at a distance of 500 mm from each support. Accordingly, 1,000 mm will be the length of the fix moment value.

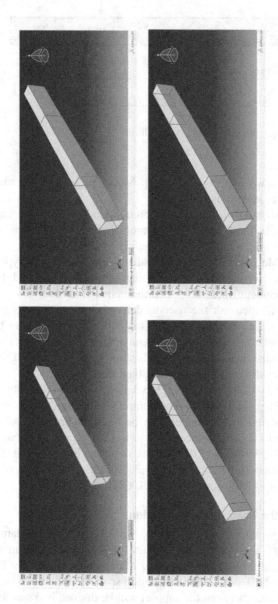

**FIGURE 5.49** Loading diagram (one half of the support span).

The loading will apply a distance to the beam at a certain location. The value of this displacement is 50 mm. In the Step Module, the displacement step has already been created. To meet the STAM Standard, four-point bending, and the displacement is applied at two locations. Each quarter of the span for each support will be exactly in the same location as for the datum plane, which is created with the R-C-Beam instance; number two and four from either direction. Assign this displacement on the 3D modeling as follows:

Create a set to assign the displacement. The displacement will be applied in two locations on the top surface of the R-C-Beam. In case the user did not create the partition and create the set in the Create Part, the user needs to do it in this step.

In this case, return to Part → select R-C-beam part → then from the vertical toolbar click on Partition Cell. Then select from the horizontal options the method of partition which is Use Datum Plane → and the program will then ask to Select the cells to partition. From the window, select the beam, and then click Done. Next, the prompt will ask the user to Select a datum plane. Choose either a datum plane; number two or four. Then the program will inform the user that the Partition definition is completed and whether to create another partition.

Repeat the same procedure as in the previous point with another datum plane. Figure 5.50 shows the procedure to create the partition.

After creating the partition on the beam, define and apply the load on the beam.

First, return to the Load Model step and again click on Create Boundary Condition but in the boundary condition dialog box the Name will be Displacement, and the Step will also be Displacement. The rest will be the same as the previous one. The Category will be Mechanical, and Types for the selected step will also be Displacement/Rotational. Click Continue.

The program will then ask the user to Select regions for the boundary condition (Create set...) → click on the two new sets which are created at the top surface of the beam → then at the Create set enter Displacement.

In the Edit Boundary Condition dialog box, review the previous input data and apply the loading → tick the box next to U2 (displacement in the Y direction) and enter -50 in the active box beside it. Then click OK to confirm.

The position and the direction of the load will appear in the model, as shown in Figure 5.51.

### 5.4.7 MESH MODULE

Meshing means dividing a geometry into smaller elements. As long as Abaqus considers the finite element method program and all the finite elements depending on one discrete element into the member and nodes, this program will divide the assembly and all parts within it into small regular or irregular elements.

To apply meshing in Abaqus, the first stage involves creating the parts. The user can immediately follow creating the different parts, start to apply the mesh for each part separated. The second stage which is recommended is to follow the same order suggested by the program to create the meshing after creating the assembly, interaction, step, and load.

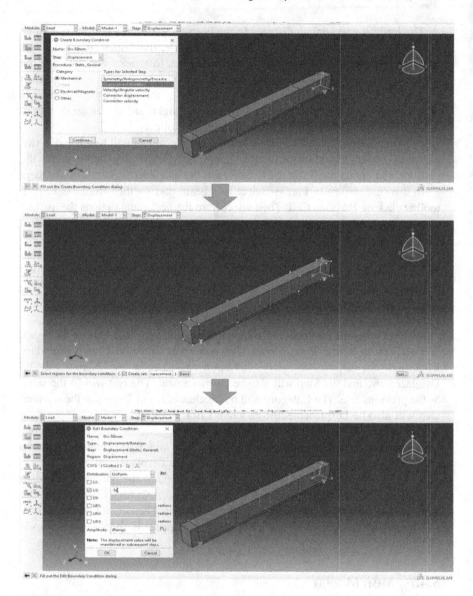

**FIGURE 5.50**   Create partition for the R-C-beam part.

The program provides some indication that will inform the user of the condition of the parts, as shown in Figure 5.52, the Mesh Controls dialog box. This box will provide information about the available Element Shape and Technique for a certain part. In general, there will be four Element Shapes that the user can select depending on the situation of the modeling. These four shapes are Hex, Hex-dominated, Tet, and Wedge. In the Technique part, it includes As Is, Free, Structured, Sweep Bottom-up, and Multiple.

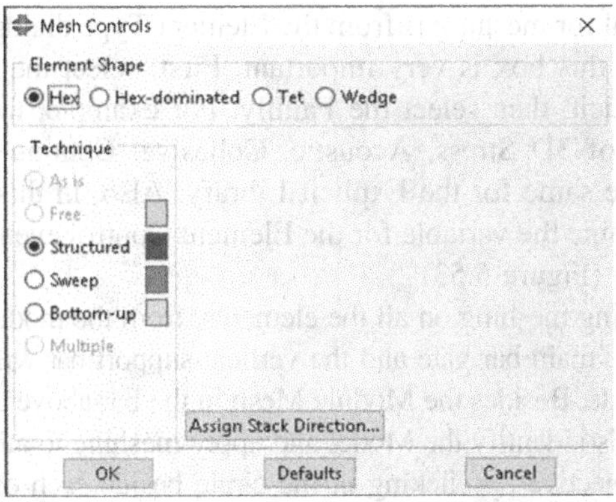

**FIGURE 5.51**  Create loading.

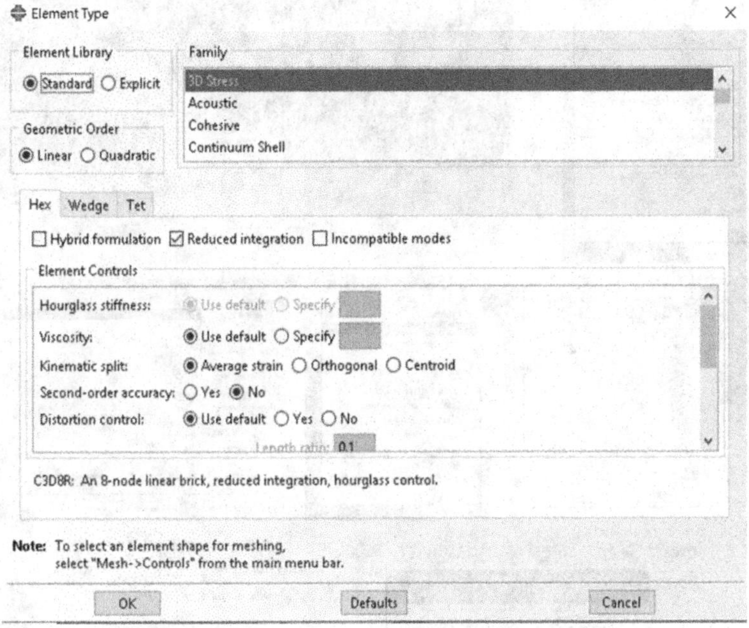

**FIGURE 5.52**  Mesh control.

Some parts are presented as a example and are ready to start applying the mesh step directly. However, some other parts consider more complicated shapes and need treatment and a partition before meshing. The best kind of meshing is when the element is under the Hex shape using the structured technique.

Another control for meshing is from the Element Type dialog box. Identifying the parameters in this box is very important. First, select the Element Library, Standard, or Explicit, then select the Family. For example, for the standard library it consists of 3D Stress, Acoustic, Cohesive, Continuum Shell, and so forth. This will be same for the Explicit Library. Also, in this dialog box, it is required to determine the variable for the Element Control even for Hex, Wedge, and the Tet shape (Figure 5.53).

To begin applying meshing on all the elements, from the module, scroll down to Module Mesh. The main bar gate and the vertical support bar will change instantly to the meshing mode. Besides the Module Mesh in the bar above the working space, the program will also identify the Model and apply meshing to all the assemblies or part of it from Objective by clicking in the circle beside each one. There are two types of elements according to the meshing. The R-C-Beam and C-Jacket will be the first group. The second group will include Steel-bar, Steel-links, and CFRP-rods.

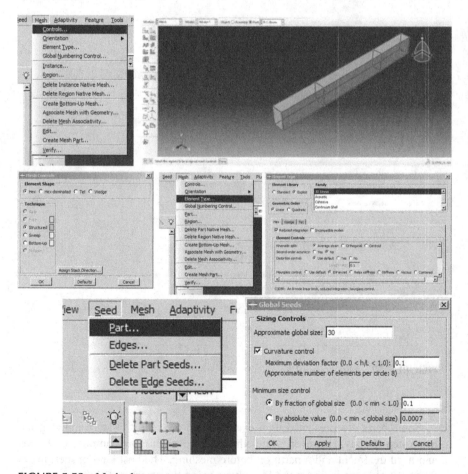

**FIGURE 5.53**  Mesh element type.

The process to apply the mesh is divided into two sequenced steps. The first step involves creating the seed for the elements after identifying the mesh control and the element type. The second step is creating the mesh and verifying the mesh. Two examples for each group will be explained.

### 5.4.7.1 R-C-Beam Meshing

From Objective click on the circle beside Parts and scroll on C-R-Beam. The beam will appear in the working area with a green color, which indicates it is a structured technique.

From the top bar, click Mesh → from the sub-Menu click on Controls.

The prompt will ask the user to Select the regions to be assigned Mesh Controls. In the working area assign the beam by creating a window around it then click Done.

A Mesh Control dialog box will appear with Hex in the Element Shape and Structured with a green color on the technique. Then, it is ready for meshing → click OK.

Again, from the top bar, click Mesh → from the sub-Menu click on Element Type.

In the prompt space, it will ask the user to Select the regions to be assigned and Element Type. In the working, assign the beam by creating a window around it → then click Done.

An element type dialog box will appear → for this beam. Select standard as Element Library and 3D Stress from the Family. Select Geometric Order as Linear and for Element Control leave it as default → click OK.

Click the Seed Part icon on the vertical supporting bar.

A Global Seed dialog box will open, and in the prompt, it will ask to Set the data using the Global Seeds dialog. Make the Approximate global size as 30, and leave the rest as is. Then click Apply to view the seeds or click OK to confirm and close the dialog box (Figure 5.54).

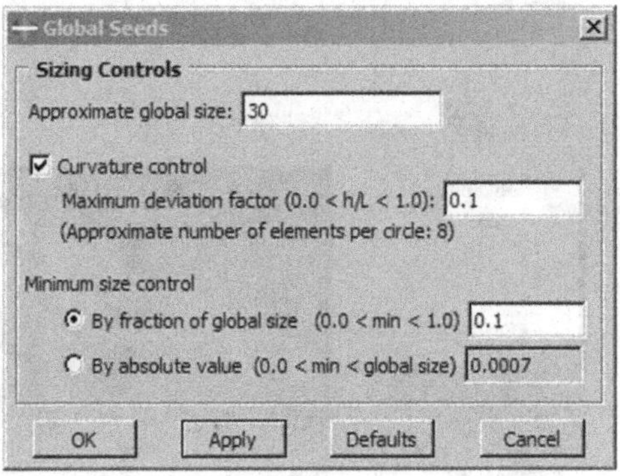

FIGURE 5.54   Control, type of element, and create seeds.

Click on the Mesh Part icon on the vertical supporting bar. In the prompt, it will ask whether it is OK to mesh the part. Yes or No → select Yes.

The program will create the mesh for this part as the approximate size of a seed, which was defined earlier (Figure 5.55).

To verify the mesh again, from the vertical supporting bar, click on Verify Mesh. Select the beam when in the prompt and Select the regions to verify (here

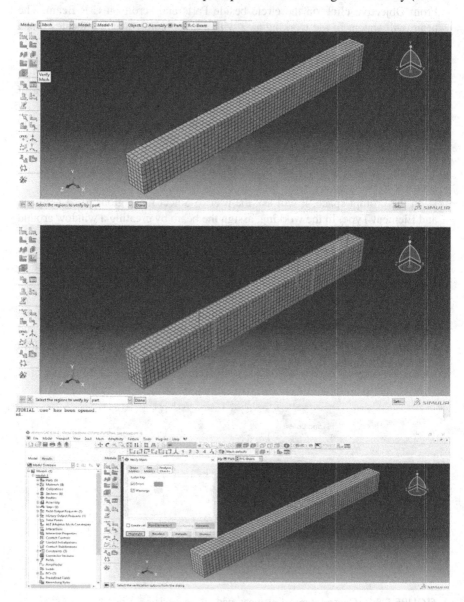

**FIGURE 5.55** Create mesh.

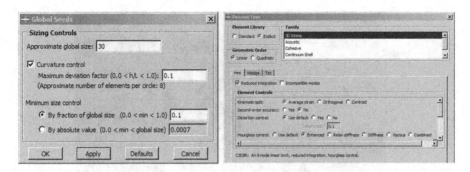

**FIGURE 5.56**   Verify mesh.

provide the level of regions part, geometricgeometric, and element) → then click Done.

The Verify Box will appear presenting three options: Shape Matrices, Size Matrices, and Analysis Checks. Go to the Analysis Checks with Highlight (with the Color Key: The Errors will be in pink color,color and the Warnings will be ain yellow color). Also, from the Message Area at the bottom read the number of errors and warnings. Here, as shown in the Figure,figure, there are 0 errors and 0 warnings. That means that the mesh is ok → click dismiss to close and move to another part for meshing (Figure 5.56).

### 5.4.7.2 Jacking Concrete

Jacketing concrete is a concrete element. The procedures applied to the previous concrete beam will be the same for jacketing. As mentioned above, this is to make the density or the size of the meshing for all elements the same. The same size of the creating seed will be equal to 30.

Figure 5.57 shows two tables; the left table is the Element Type dialog box of jacketing concrete, and the other one is the Creating Seed dialog box.

### 5.4.7.3 Steel Links

All the steel members, Steel-bar and Steel-links are 2D members as long as they are created as Wire in the 2D Planner and they are defined as Truss element material having a certain thickness. However, the meshing type for these parts will be marginally different. In the Element Type dialog box, as shown in Figure 5.58, the family here will be selected as Truss.

Again, when creating the Global Seed from the Global Seed dialog box, the Approximate global seed will be taken as 20. Then, creating and Verifying the meshing is done using the same tools used in case of the Concrete Beams.

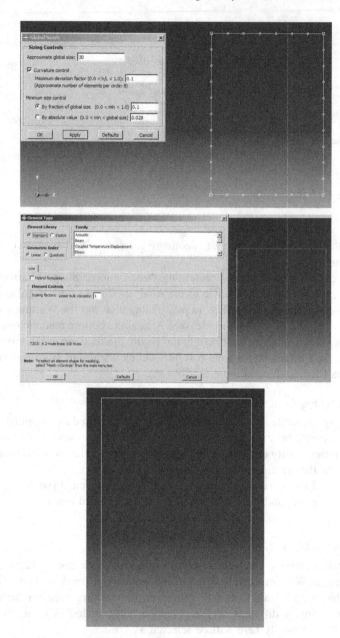

**FIGURE 5.57** Meshing of concrete jacketing.

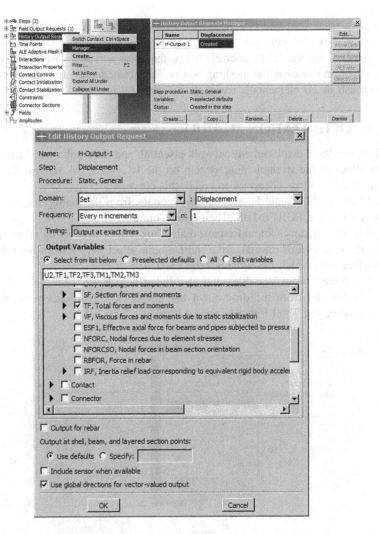

**FIGURE 5.58** Meshing of steel-links.

### 5.4.7.4 Steel-Bar and CFRP Rods

The Steel-bar part and even the CFRP Rods part both are the same as the Steel-link, which is Wire in the 2D Planner Elements. The only difference between them is the material assigned to each. So, the same steps will be followed.

Indeed, it is very important to notice here that in both the beam and jacketing concretes the color of the elements at the beginning of creating the meshing is a Green color which indicates that it is structured and the meshing can be applied. If the color is yellow or dark brown, the situation would be different. In this case, the program cannot create meshing directly without treatment that is done by the user, such as partitioning the elements into smaller elements and so forth.

Second, for the 2D Planner elements, the color was near to Pink (Magenta); however, in this case, the meshing was created immediately. In general, creating the mesh in 2D elements is easier than creating it in a solid 3D element.

## 5.5  OPTIMIZATION MODULE

Here, the force-displacement curve is essential, and the results will help explain the effect of the strengthening system on the RC beam. So, request the output data which will be necessary to create this curve.

Click on the Create History Output icon → from the dialog box create and enter the name FORCE-DIS. Then click Continue.

The Edit History Output Request dialog box will open showing many options. For the case of this tutorial, from the Domain select Set → from Set select the Displacement which has already been created.

From the Output Variable, click the dialog box. Besides the Select from the list below and from the drop-down list just click U2 displacement from the Translation and Total Force as shown in Figure 5.59 below.

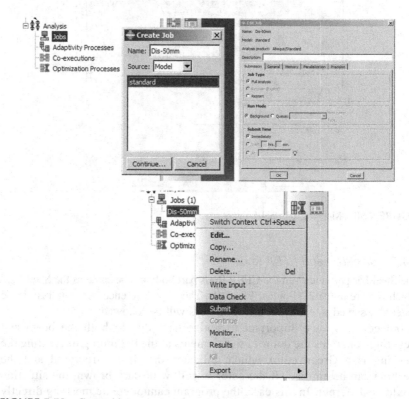

**FIGURE 5.59**  Create history output.

## 5.6   ANALYSIS: JOB MODULE

Although the Job module is considered to be one of the easiest steps among all the steps in the modeling procedures, this step is very important because the modeling will be transferred from the preprocess stage into the process stage through this step. Further, this step is considered as a station for checking all previous steps starting from creating the parts until creating the meshing. If there are any missing or incorrect data, it will be identified in this step. So, the value of this step is in relation to checking and reviewing all previous steps.

The program divides the mistakes into two categories. The first category under the name Warnings provides information regarding incorrect data or any incompatibility with regard to these mistakes. However, in the other category under Errors the mistakes are serious and are relating to the input data, as a result of which the program will stop further analysis, providing the user the chance to modify or correct the input files before the program can continue analyzing the data again.

To transfer from the mesh module or optimization module to the Job module, and run the job analysis, use the same strategy as follows:

From the module in the menu bar above the working area, this time scroll through the list until the Job module. The main top menu bar and the vertical supporting tools will change immediately to the job step.

At the top of the vertical supporting bar click on Create Job → a Create Job dialog box will appear → in this box, it will ask for the Name of the job (for example, DIS-50mm) and the Source, either the model or input file. Keep it as Model 1 and then click Continue.

An Edit Job dialog box will open. In the Description dialog box enter Displacement → for the rest of the variables do not make any changes but keep them as the default → then click OK and the dialog box will close.

Next, click on the icon besides it, which is Job Manager → a Job Manager dialog box will open with the Name of the created job and its Resource. Type of analysis is full analysis, and the status is None.

In this box, there are two categories. The first category at the top will contain Write Input, Data Check, Submit, Continue, Monitor, Results, Kill, and finally Dismiss which follow the order in performing the job analysis. Next, the bottom category includes Create, Edit, Copy, Rename, and Delete; all of them related to the analysis of Creating the Job.

Start the analysis with the order from the right [vertical order] from the top → click on Write Input → the program will start to generate the data or give a message regarding incorrect data to fix (Figure 5.60).

Here the message concerning the constraint which was created previously in the instruction step has some problems; either the region is deleted or renamed or suppressed. This leads to the need for checking the constraint region. In the message, it will ask to Continue Without Correcting the date or NO. Click on NO as it is compulsory to continue generating the data.

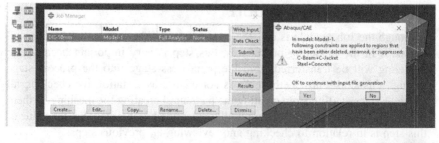

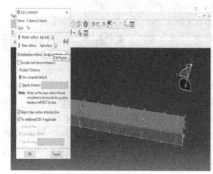
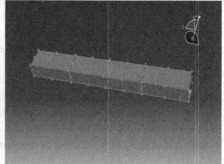

**FIGURE 5.60** Create a job.

Return to the Interaction Step and click on Constraint Manager → the Constraint Manager dialog box will open → start to correct the problematic constraint.

As mentioned in the message, the problem comes from two constraints C-Beam + C-Jacket and Steel-Concrete. Both are related to the R-C-Beam. This error comes from the region that is already defined and both constraints have been deleted already as a result of the partition being applied to the R-C-Beam in the load step in applying the loading.

The Partition deleted the old part R-C-Beam and all its sets and surface. A new element named C-R-Beam-1 in the assembly will therefore be needed to be created.

From the Constraint Manager dialog box double click the C-Beam + C-Jacket → in the message dialog box, Click Yes to continue editing the interaction and redefine the regions for this constraint. In the Edit Constraint box click on "raw" beside the Host region and by using the Assembly Display Option again select the R-C-Beam-1 → and create a new set with the name host beam and click Done to confirm.

Repeat the same step for the second error constraint. From the Constraint Manager dialog box double click on Steel-Concrete → in the message dialog box, click on Yes to continue editing the interaction and redefine the regions for this constraint. In the Edit Constraint box click on "raw" beside Master surface. Using the Assembly Display Option again and View Manipulation select all the surface on the bottom of R-C-Beam-1 and create a new set named top-side and click Done to confirm, as shown in Figure 5.61.

After fixing the constraint missing data, return to the Job Manager dialog box. Again, click on Write Data and in the message box click OK to allow over-writing on DIS-50mm → the status of the job will change to Check Submitted.

Click on Monitor to follow the analysis situation. The Monitor dialog box will be opened. In this box, the most important parameters are Status, Warnings, and Errors.

If the status changed to Aborted, that means there is an error which has caused the analysis to stop. Open the Monitor Box to discover the errors. If there is an error 2920, it means the elements have missing property definitions.

This error is common from neglecting to assign the material to one or more of the parts or mistakes in creating the mesh for the different parts. As long as a partition was applied for the R-C-Beam part that means the previous section assigned on it has already been deleted.

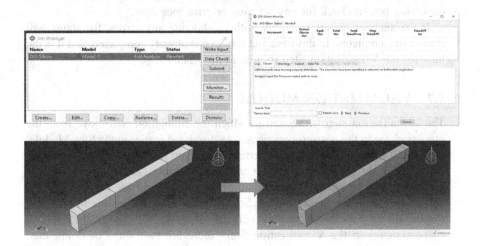

**FIGURE 5.61**    Correct the constraint error.

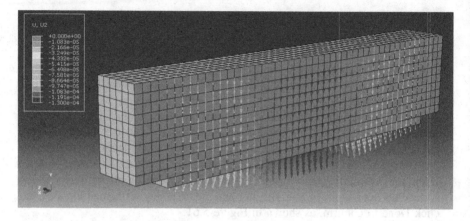

**FIGURE 5.62**   Redefine the materials for the RC beam.

Go to the Property Step. Here the user will find the R-C-Beam with a gray color, which means no materials defined → reassign the section as a Concrete-beam section as shown in Figure 5.62.

Next, click on Write Data on the job manager and the status after some time will be changed to completed. The job input file has been written to *DIS-50mm.inp* in the comment area.

Then, click on the Data Check.

Finally, click on submit for analysis. This stage will take time according to numerous factors as mentioned previously, such as the size and type of the assembly besides the setting for step, meshing, loading, and other factors.

During the period of analysis, follow the processing status from monitoring the dialog box to check for any warning or error messages.

If the analysis moves smoothly without showing any errors, the status will change to Completed. In this case, the Job Step is already completed, and it is the time for the Visualization Step to present the results.

## 5.7   VISUALIZATION MODULE

The Visualization step is the step in which the simulation will move from the process phase to the post-process phase. As discussed earlier in the Step Module, the request for specific results at a certain place even set, point or even surface is controlled in the Create Field Output and Create History Output. In case there is no specific request for a certain result, the program will provide the user with the default results in the whole model. So, with that in mind, in this tutorial, the results will be presented as general defaults.

To transfer to the Visualization Step, there are two ways of doing this. First, after finishing the analysis in the previous step from the job manager dialog box, click on results directly, and the program will transfer to the

Visualization Step. Alternatively, another way is by scrolling to the Visualization Step from the list besides the module. When the Visualization Step becomes active, all the other bars in the window, even the left window, will completely change from the model face into the result face for the first time since the start of the work.

The Field Output Bar will show the results, which are Primary, Deformed, and Symbol. For each result, parameters will appear in the modeling.

To display the maximum and minimum value for each parameter in the main window, from the main bar click Viewport → from the sub-menu click on the Viewport Annotation Option.

Select the Legend gate and from the Viewport Annotation Option → click on the empty dialog box beside Show min/max values → click on Apply and OK to confirm and exit (Figure 5.63).

The location of the max/min value in the graphical modeling can be determined by the following:

From the vertical supporting bar click on the Contour Option icon → the Contour Plot Option dialog box opens (Figure 5.64).

From this dialog box → go to the Limits gate → click on the empty dialog boxes beside Show Location for both the max and min values. → Click on Apply and then OK to confirm and move to another order.

## 5.8 ANALYSIS RESULT

The stresses can be shown in the primary model by selecting the Primary Model from the Filed Output Bar then select stress from the window besides it. From the list sliding window, the user can choose the stresses in a different principle direction X, Y, and Z and also the secondary direction.

Figure 5.65 shows the contour value of the stress S33 in the whole model. Also, in the small box at the top right corner, the stress values will appear in different contour colors. Furthermore, at the bottom, the maximum values can be seen and also the location at the top-middle portion of the beam. However, the minimum stress value will be shown in the support line.

Figure 5.66 shows the value and distribution of the stress for the different parts individually inside the module.

Energy dissipation is another important output result. The energy will reflect the amount of energy needed to deform the modeling. The energy can be an illustration in the graph connected to the energy dissipation variable with time. To draw this curve, follow the steps below:

From the top bar click on Result → from the sub-menu click on History Output.

A History Output dialog box will open → this box shows many Variable Output results → scroll using the mouse and select Total Energy for the Output set: E Total for the Whole Model → then click Plot to draw it in the main window, as shown in Figure 5.67.

Also, in the results output the user can show the displacement for the whole model under the loading by using U displacement in the History Output. Select

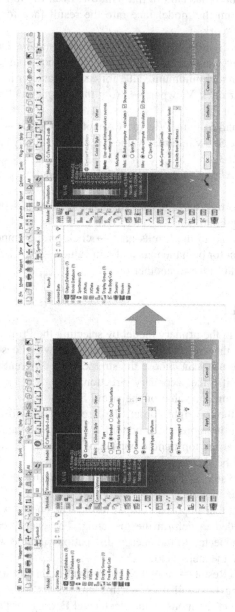

**FIGURE 5.63** Viewport annotation options.

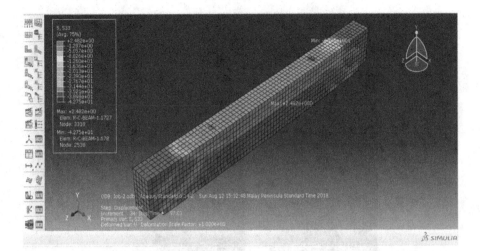

**FIGURE 5.64** Contour plot options.

the displacement in any direction U1, U2, and U3 in X, Y, and Z, respectively, as shown in Figure 5.68.

For all previous variables, the program can provide many ways to show the tendency or how these parameters change with increase in time or increment. To apply this, select from the main bar or the vertical supporting bar, and then click on the Animate: Scale Factor, the Animate: Time History or the Animate: Harmonic icons. The program provides the user the flexibility to control and modify these Animates, as shown in Figure 5.69.

The increase in the strength capacity of the strengthened beam is expressed in terms of the Force-Displacement Curve. Also, by using the same tool, History Output, the program can easily draw the change in the displacement with time and also the force with time. Furthermore, the program can create a new curve with whatever parameters are selected. To create a force-displacement curve, apply the following steps:

From the vertical supporting bar click in the Create XY Date → the create XY Data dialog box will appear and ask the user to select the Source for the data → click on ODP History Output.

Again, the History Output dialog box will be open and ask the user to select the Output Variable data to plot in the program.

For example, click on Total Force Tf2 P1... → then click Plot → on the screen, the program will draw the curve of the time and displacement in the selected set.

Click on Save As → a Save As dialog box will open asking the user to provide a Name to the Save Operation → enter the name FORCE-N and click on avg ((XY, XY...)) → Click the box besides the Plot curve → then click OK. The average curve for displacement with time will be drawn in the main window, as shown in Figure 5.70.

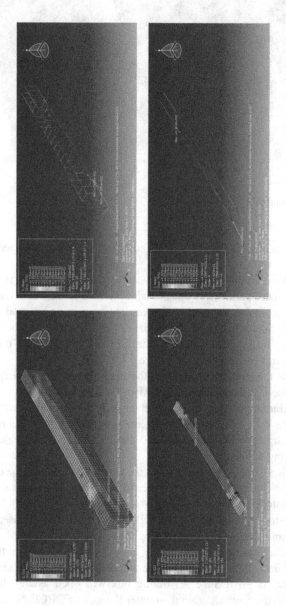

**FIGURE 5.65** Distribution of the stress in the whole model.

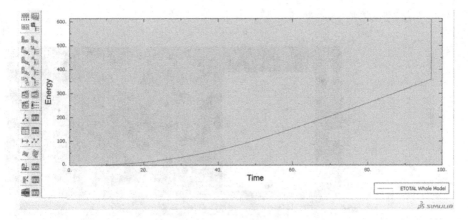

**FIGURE 5.66** Distribution of stress for different parts.

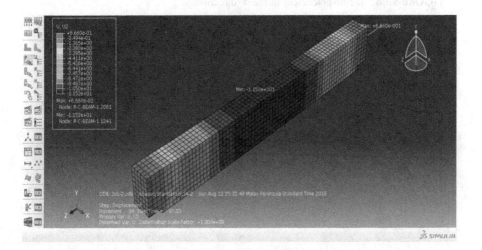

**FIGURE 5.67** ETotal in the whole model.

The same step is applied as in the previous step but here, change the selection from Special displacement to Special displacement U2... the name of the average total force curve will be DIS-mm, as shown in Figure 5.71.

To draw the Force-Displacement curve, which is the most important curve to express the behavior of the strengthened beam, again, click on the Create XY Data icon and then click on Operate on XY Data.

The Operate on XY Data dialog box will open → from the right, select the Operation as Combine (X, X) → then click on Save As under FORCE-DIS. Finally, click on Plot Expression to display the curve (Figure 5.72).

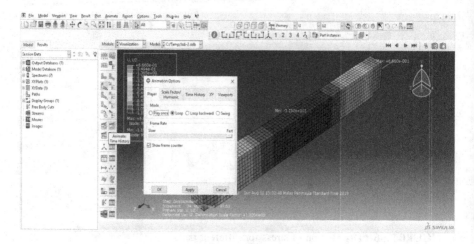

**FIGURE 5.68** U2 displacement in the Y direction.

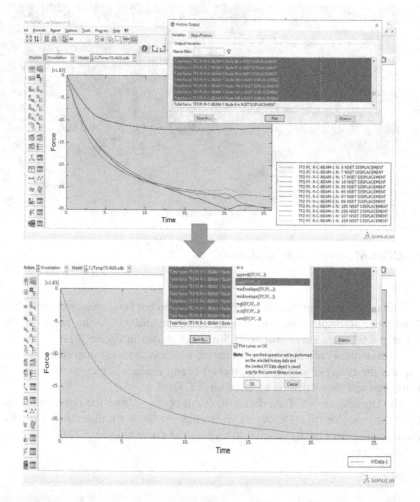

**FIGURE 5.69** Animation option.

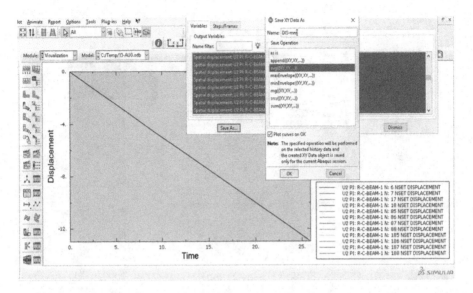

**FIGURE 5.70**  force-time curve.

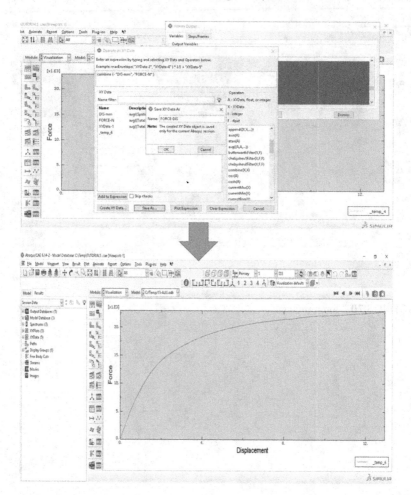

**FIGURE 5.71**   displacement-time curve.

# 6 Explosive Loading on a Sandwich Plate

## 6.1  INTRODUCTION

One of the challenging tasks for a structure designer is to predict the structural response against the explosion force and the loads caused by shocks. The occurrence of an air blast causes a compressive air face that hits the structures. The structures absorb some of this energy which increases tensional force and mechanical strain on the structural members. Depending on the amount of energy absorbed by the structures, some of the structural members experience of the plastic deformation or even permanent damages.

## 6.2  PROBLEM DESCRIPTION

This example evaluates the effects of an explosion on a sandwich plate and its deformation. The sandwich plate which can be considered as a wall or ceiling is composed of two layers of concrete with a insulation sheet in between (which can be air). For this purpose, a sandwich plate with an external dimension of 1,200 × 1,200 mm and a total thickness of 60 mm is considered. An explosive mass of 0.5 kg TNT is placed at the height of 100 mm from the center of its upper plate. The dimensions and specifications of the panel sandwich are shown in Figure 6.1.

Air in the middle portion of the sandwich panels may be considered as an ideal insulator for some structures. However, on the other hand, the stiffness would be relatively low to mid range, which makes the wall panel vulnerable, especially against out of plane loads such as shocking and sudden loads due to explosion charges.

The sandwich plate is made of steel material and its plasticity behavior depends on the applied strain rate. Whereas, in high-velocity forces, such as an explosion, the material deforms in varying rates.

## 6.3  OBJECTIVES

1. To study the response of sandwich panel to the explosion force
2. To investigate about the energy variation and applied strain rate on the sandwich panel

DOI: 10.1201/9781003213369-6

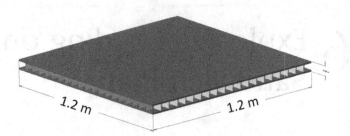

**FIGURE 6.1**   Sandwich plate dimensions.

## 6.4   MODELING

Run Abaqus/CAE from the start menu and close the StartSession dialog box (Figure 6.2).

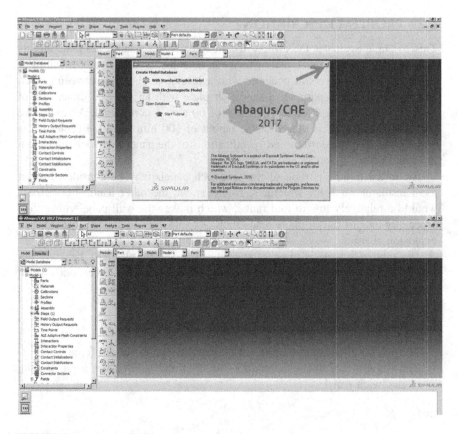

**FIGURE 6.2**   Abaqus/CAE.

## 6.4.1 Part Module

At first, the geometry should be defined. In this problem, the geometry consists of two main sheets and a stiffener area that should be individually drawn and placed in the appropriate position. The plate will be considered as solid and stiffeners as a shell.

For the simplicity of work, and due to the geometric symmetry and circumferential condition in this problem, it was attempted to model one-quarter of it and generalize the whole.

Double click on Parts in the ModelTree to open the CreatePart dialog box. Name the part as Plate and set the Approximate size as 2. Do not change any of the other defaults and click Continue to open Sketcher.

Select Create Lines: [Rectangle] and draw a rectangle.

Select AddDimension and determine the plate dimensions as 0.6 x 0.6 square.

Click the middle mouse button twice to open the EditBaseExtrusion dialog box and enter 0.01 as the Depth. Click OK to apply, and the dialog box will be closed (Figure 6.3).

### 6.4.1.1 Create Stiffener

Double click on Parts in the ModelTree to open the CreatePart dialog box. Name the part as Stiffener, then set the shape as Shell and choose the Extrusion as type. Finally, set the Approximate size as 2. Then, click Continue to open Sketcher.

Select Create Lines: [Connected]and draw a vertical line.

Select AddDimension and consider the length 0.6.

Select LinearPattern and select the line and click the middle mouse button to open the LinearPattern dialog box. Consider the pattern just in Direction 1 and set the Number and Spacing as 9 and 0.06, respectively, and click OK to make an array of the line.

Draw a horizontal line and determine the length as 0.6 and distance to the vertical line as 0.06.

Define the length as 0.6.

Repeat the pattern for Direction 2 and keep the Number and Spacing as mentioned above.

Click the middle mouse button twice to open EditBaseExtrusion and enter 0.04 as the Depth. Then click OK to apply and close (Figure 6.4).

## 6.4.2 Material Properties

The sandwich plate is completely made of steel, and its properties are enlisted in Table 6.1 and Table 6.2.

One of the complications caused by an explosion on structures is the severe amount of damage caused and plastic behavior. Hence, in the material properties of this model, elastic and plastic behavior along with damage parameters should be considered.

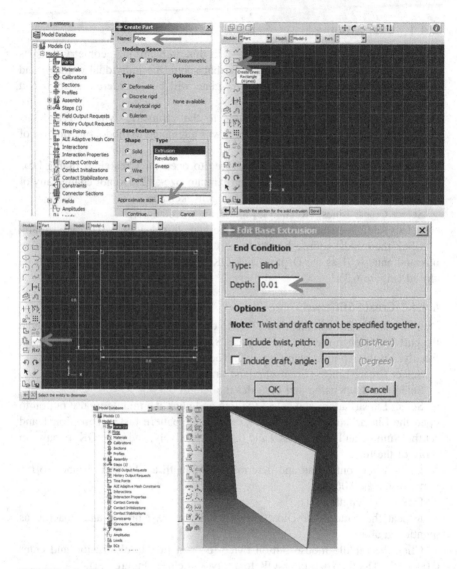

**FIGURE 6.3** Creating the plate.

For damage modeling in soft metals (such as for steel), ductile damage should be defined, which is one of the progressive failures. In this method, damage based on the strain of plastic will be obtained as given in Equation (6.1).

$$D = \bar{\varepsilon}_D^{pl}(\eta \bar{\varepsilon}_D^{pl}) \tag{6.1}$$

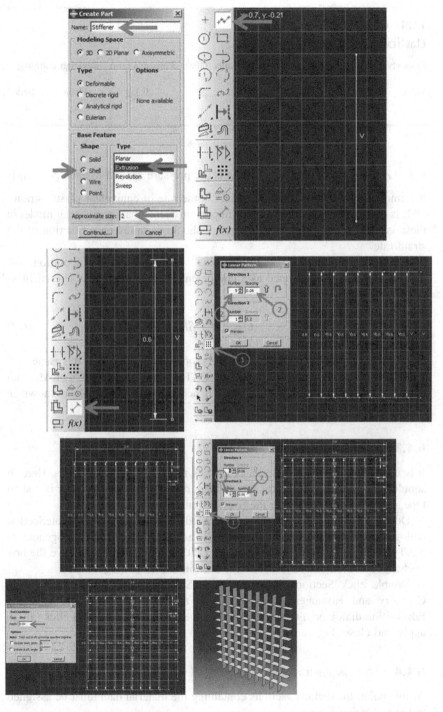

**FIGURE 6.4** Creating stiffener.

**TABLE 6.1**

**Elasticity and damping properties**

| Property | Density (kg/m³) | Young's modulus (GPa) | Poisson's ratio | Rayleigh damping | |
|---|---|---|---|---|---|
| Value | 7850 | 200 | 0.3 | Alpha | Beta |
| | | | | 50 | 0 |

where $\eta = -\frac{p}{q}$ stress triaxiality, and $p$ is the pressure applied to each point, $q$ is the amount of Von Mises stress, and $\dot{\bar{\varepsilon}}^p$ is the rate of equivalent plastic strain.

It is essential to note this point in impact problems that the material model in these issues for plastic areas and damage should certainly be a function of the strain rate.

Also, for damping, Rayleigh damping was used. In this definition, the damping matrix is determined as a coefficient of the mass matrix and stiffness matrix (Equation (6.2)).

$$C = \alpha M + \beta K \qquad (6.2)$$

Hence, this property is considered in the properties provided for this issue.

Double click on Material in the ModelTree and open the CreateMaterial dialog box. Define the material properties according to Table 6.1, as shown in Figure 6.5.

### 6.4.3 SECTIONS PROPERTIES

It is necessary to put the material properties in corresponding sections. Here, it should be noted that two pieces are defined: one is solid and the other is a shell type. It is necessary to define the two sections.

Double click on the Sections in the ModelTree to open the CreateSection dialog box and name it as Plate. Leave Solid as the Category and Homogenous as the Type and click Continue to open the EditSection dialog box. Leave the box unchanged and click OK to define the section and close (Figure 6.6)

Double click Sections and name it as Stiffener, then select Shell as the Category and Homogenous as the Type and click Continue to open the EditSection dialog box. Consider 0.001 as the Thickness value and click OK to apply and close (Figure 6.7).

### 6.4.4 SECTION ASSIGNMENT

At this point, the defined sections containing the material data must be assigned to their associated parts.

**TABLE 6.2**
**Plasticity and damage properties**

| | Plasticity | | | | | Ductile damage | | |
|---|---|---|---|---|---|---|---|---|
| Yield stress | Plastic strain | Strain rate | Yield stress | Plastic strain | Strain rate | Fracture strain | Stress triaxiality | Strain rate |
| 776,000,000 | 0 | 0 | 808,000,000 | 0 | 1 | 2.31 | -3.33 | 0.001 |
| 809,000,000 | 0.01 | 0 | 842,000,000 | 0.01 | 1 | 2.31 | -0.333 | 0.001 |
| 829,000,000 | 0.02 | 0 | 869,000,000 | 0.02 | 1 | 2.18 | -0.267 | 0.001 |
| 842,000,000 | 0.03 | 0 | 893,000,000 | 0.03 | 1 | 2.06 | -0.2 | 0.001 |
| 866,000,000 | 0.06 | 0 | 946,000,000 | 0.06 | 1 | 1.95 | -0.133 | 0.001 |
| 883,000,000 | 0.1 | 0 | 998,000,000 | 0.1 | 1 | 1.85 | -0.0667 | 0.001 |
| 895,000,000 | 0.15 | 0 | 1,050,000,000 | 0.15 | 1 | 1.76 | 0 | 0.001 |
| 910,000,000 | 0.25 | 0 | 1,120,000,000 | 0.25 | 1 | 1.67 | 0.0667 | 0.001 |
| 922,000,000 | 0.4 | 0 | 1,190,000,000 | 0.4 | 1 | 1.59 | 0.133 | 0.001 |
| 953,000,000 | 2 | 0 | 1,490,000,000 | 2 | 1 | 1.52 | 0.2 | 0.001 |
| 791,000,000 | 0 | 0.001 | 810,000,000 | 0 | 10 | 1.46 | 0.267 | 0.001 |
| 824,000,000 | 0.01 | 0.001 | 846,000,000 | 0.01 | 10 | 1.4 | 0.333 | 0.001 |
| 846,000,000 | 0.02 | 0.001 | 876,000,000 | 0.02 | 10 | 1.35 | 0.4 | 0.001 |
| 863,000,000 | 0.03 | 0.001 | 901,000,000 | 0.03 | 10 | 1.3 | 0.467 | 0.001 |
| 899,000,000 | 0.06 | 0.001 | 960,000,000 | 0.06 | 10 | 1.26 | 0.533 | 0.001 |
| 931,000,000 | 0.1 | 0.001 | 1,020,000,000 | 0.1 | 10 | 1.23 | 0.6 | 0.001 |
| 958,000,000 | 0.15 | 0.001 | 1,070,000,000 | 0.15 | 10 | 1.2 | 0.667 | 0.001 |
| 995,000,000 | 0.25 | 0.001 | 1,150,000,000 | 0.25 | 10 | 1.15 | 0.73 | 0.001 |
| 1,030,000,000 | 0.4 | 0.001 | 1,240,000,000 | 0.4 | 10 | 1.06 | 0.851 | 0.001 |
| 1,170,000,000 | 2 | 0.001 | 1,600,000,000 | 2 | 10 | 0.945 | 1.02 | 0.001 |
| 799,000,000 | 0 | 0.01 | 812,000,000 | 0 | 100 | 0.816 | 1.24 | 0.001 |
| 831,000,000 | 0.01 | 0.01 | 850,000,000 | 0.01 | 100 | 0.685 | 1.51 | 0.001 |
| 855,000,000 | 0.02 | 0.01 | 882,000,000 | 0.02 | 100 | 0.202 | 3.33 | 0.001 |

*(Continued)*

**TABLE 6.2 (Continued)**
**Plasticity and damage properties**

| Plasticity | | | | | | Ductile damage | | |
|---|---|---|---|---|---|---|---|---|
| Yield stress | Plastic strain | Strain rate | Yield stress | Plastic strain | Strain rate | Fracture strain | Stress triaxiality | Strain rate |
| 874,000,000 | 0.03 | 0.01 | 909,000,000 | 0.03 | 100 | 2.31 | -3.33 | 0.001 |
| 916,000,000 | 0.06 | 0.01 | 974,000,000 | 0.06 | 100 | 2.31 | -0.333 | 0.001 |
| 955,000,000 | 0.1 | 0.01 | 1,040,000,000 | 0.1 | 100 | 2.18 | -0.267 | 0.001 |
| 989,000,000 | 0.15 | 0.01 | 1,100,000,000 | 0.15 | 100 | 2.06 | -0.2 | 0.001 |
| 1,040,000,000 | 0.25 | 0.01 | 1,190,000,000 | 0.25 | 100 | 1.95 | -0.133 | 0.001 |
| 1,090,000,000 | 0.4 | 0.01 | 1,280,000,000 | 0.4 | 100 | 1.85 | -0.0667 | 0.001 |
| 1,280,000,000 | 2 | 0.01 | 1,700,000,000 | 2 | 100 | 1.76 | 0 | 0.001 |
| 805,000,000 | 0 | 0.1 | 815,000,000 | 0 | 1,000 | 1.67 | 0.0667 | 0.001 |
| 838,000,000 | 0.01 | 0.1 | 855,000,000 | 0.01 | 1,000 | 1.59 | 0.133 | 0.001 |
| 863,000,000 | 0.02 | 0.1 | 888,000,000 | 0.02 | 1,000 | 1.52 | 0.2 | 0.001 |
| 884,000,000 | 0.03 | 0.1 | 917,000,000 | 0.03 | 1,000 | 1.46 | 0.267 | 0.001 |
| 933,000,000 | 0.06 | 0.1 | 987,000,000 | 0.06 | 1,000 | 1.4 | 0.333 | 0.001 |
| 978,000,000 | 0.1 | 0.1 | 1,060,000,000 | 0.1 | 1,000 | 1.35 | 0.4 | 0.001 |
| 1,020,000,000 | 0.15 | 0.1 | 1,130,000,000 | 0.15 | 1,000 | 1.3 | 0.467 | 0.001 |
| 1,080,000,000 | 0.25 | 0.1 | 1,230,000,000 | 0.25 | 1,000 | 1.26 | 0.533 | 0.001 |
| 1,140,000,000 | 0.4 | 0.1 | 1,330,000,000 | 0.4 | 1,000 | 1.23 | 0.6 | 0.001 |
| 1,390,000,000 | 2 | 0.1 | 1,810,000,000 | 2 | 1,000 | 1.2 | 0.667 | 0.001 |
| | | | | | | 1.15 | 0.73 | 0.001 |
| | | | | | | 1.06 | 0.851 | 0.001 |
| | | | | | | 0.945 | 1.02 | 0.001 |
| | | | | | | 0.816 | 1.24 | 0.001 |
| | | | | | | 0.685 | 1.51 | 0.001 |
| | | | | | | 0.202 | 3.33 | 0.001 |

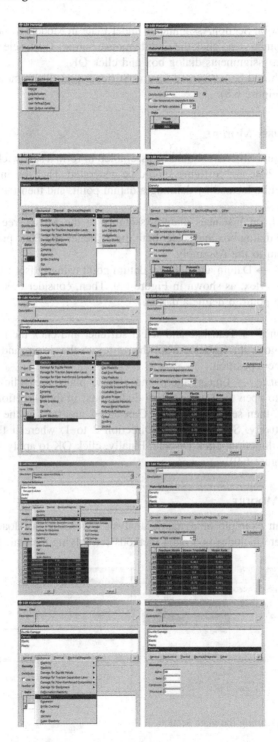

**FIGURE 6.5** Defining the material model.

Double click on SectionAssignments under Plate in ModelTree and select the entire plate geometry and click the middle mouse button. Then, select the Plate in the EditSectionAssignments dialog box and click OK.

The same should be performed for the Stiffener part; Plate and stiffener will be recolored (Figure 6.8).

### 6.4.5 ASSEMBLY MODULE

Both the Plate and Stiffener should be assembled to form a sandwich plate. First, an instance of Plate and Stiffener should be generated. Then, an instance of the stiffener should be located by defining a datum point, and then the plate should be copied.

Double click on Instances under Assembly in the ModelTree to open the CreateInstance dialog box and select both the Stiffener and Plate parts, and click OK to add them to the Assembly.

Select Tools → Datum and define a datum point using Offset from the point, and select the vertex, as shown in Figure 6.9. Then, consider 0.06, 0, 0 for the offset from the point to define a datum.

Figure 6.9

Select Instance → Translate and select stiffener and click the middle mouse button. Then move the instance from the vertex to the datum and click OK to apply for the new position.

Select Instance → LinearPattern and choose plate, then click the middle mouse button to open the LinearPattern dialog box. In the Direction1 box, click Direction icon, then select the Z axis. Consider 2 and 0.05 as the Number and Offset, respectively. Set Number in Direction 2 to 1, where it then makes a pattern. In the direction, click Disable. Finally, click OK to apply and close the dialog box (Figure 6.9).

### 6.4.6 STEP MODULE

This case is an example of dynamic analysis given the fast loading change. Abaqus software includes two main methods for solving dynamic problems.

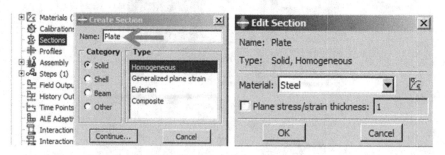

**FIGURE 6.6**  Defining the plate section.

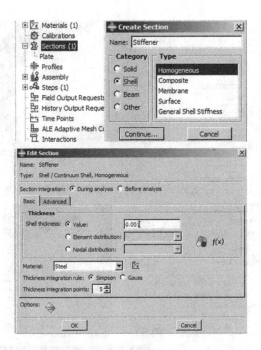

**FIGURE 6.7** Defining the stiffener section.

### 6.4.6.1 Implicit Dynamics

In this method, which is used in Abaqus/Standard solver software, the Newton-Raphson process, and simultaneous solving equations in each increment are used to analyze the problem. The speed of the analysis of these calculations highly depends on the hardware resources available, although the exact solving of the equations involves high accuracy.

### 6.4.6.2 Explicit Dynamic

In this method, which is used in Abaqus/Explicit solver software, the equations are solved sequentially based on the central difference method, in which the values of the variables in time $t + \Delta t$ are obtained based on the values at time $t$.

An explosion is a type of problem that can be simulated with explicit dynamics and is included in Abaqus/Explicit. An explosion takes place in a very short time. So the loading time period here was considered as 0.1 seconds, which involved the analysis by large displacements due to a large force that is produced by the explosion. Note that large displacements mentioned by Nlgeom are NonlinearGeometry in Abaqus.

Double click Steps in the ModelTree and select Dynamic, Explicit from the create step dialog box and click Continue to open the EditStep dialog box.

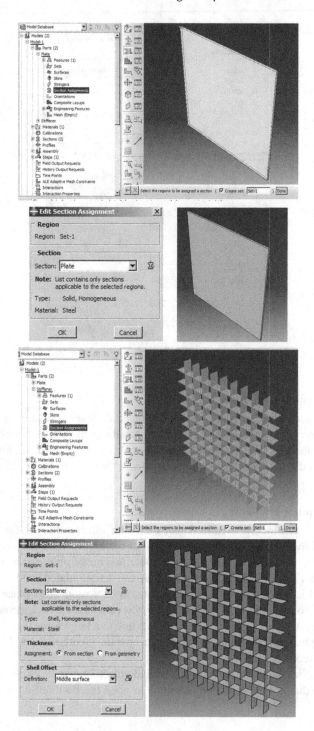

**FIGURE 6.8**  Assigning the section.

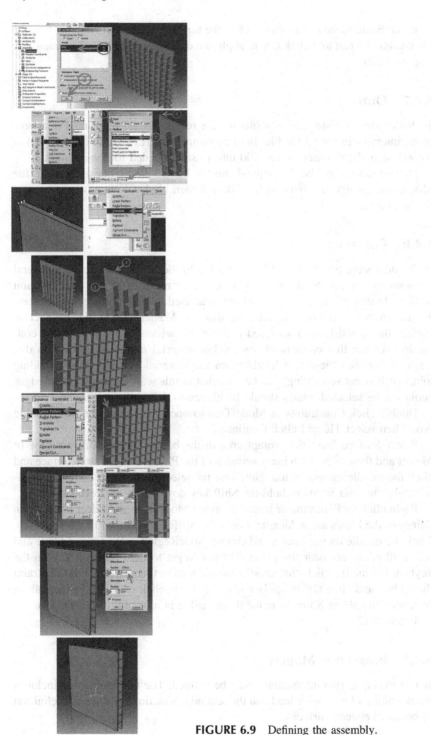

**FIGURE 6.9** Defining the assembly.

In the Basic tabbed page, enter 0.1 as the time period and set Nlgeom: On. Do not change the rest and click OK to apply the definition and close the dialog box (Figure 6.10).

### 6.4.7 OUTPUTS

To better view the simulation solutions, the results should be printed in appropriate intervals in the *.Odb file. In this example, the results will be recorded in 0.0005 seconds; in other words, 200 intervals are needed for the solution.

Double click on the F-Output-1 under FieldOutputRequests and in the ModelTree change the Interval to 200, and then click OK to save the change (Figure 6.11).

### 6.4.8 CONSTRAINT

In the next step, plates should be connected to the Stiffener. There are several methods in Abaqus to define it. One of these methods is the tie constraint method. In this method, components are connected by determining two surfaces: Master surface and Slave surface, so that the Master type is the connection surface that consists of a stiffened material model and Slave type is the connection surface that consists of a smoother material model. There are no differences between these two levels when the material is same. Many welding joints (if it is not separating) can be modeled in this way. As the stiffener edges could not be selected, plates should be hidden to define the constraint.

Double click Constraints in ModelTree to open the DefineConstraint dialog box. Then select Tie and click Continue.

Select Surface from the prompt area in the bottom of the software type of Master and then select both inner surfaces of the Plates as the Master surface and click the middle mouse button. Note that for selecting multiple components alternately, the user needs to hold the Shift key down from the keyboard.

Right click on Plates under Instances in the ModelTree and select Hide. Upon doing so, the Plates are hidden, and only the Stiffener is shown in the assembly. Click the middle mouse button and choose NodeRegion as the type of Slave and select all edges are near the plates. Do not forget to hold the shift key on the keyboard. Finally, click the middle mouse button to open the EditConstraint dialog box and click OK to apply and close. Right click on plates underneath of Instances and select Show to make them visible in assembly (Figure 6.12).

Figure 6.12

### 6.4.9 INTERACTION MODULE

In this example, two interactions must be defined. The first interaction includes the definition of explosive load and the second interaction involves the definition of contact between surfaces.

### 6.4.9.1 Explosion

The explosion in Abaqus can be performed in many ways. The explosion in the air is called CONWEP, and should be considered. The air blast loading in Abaqus is defined on a reference point. In fact, the explosion will start spherically from the point. As mentioned in the problem description, the location of the explosive load is at a distance of 100 mm from the middle of the sandwich plate, while because of the symmetry, the plates are defined as one-quarter of the original model. Here, the explosive load will be at a distance of 100 mm from the corner of one plate. The point should be defined as the explosion reference point and explosion mass considered as 0.5 kg TNT and to explode once the analysis begins.

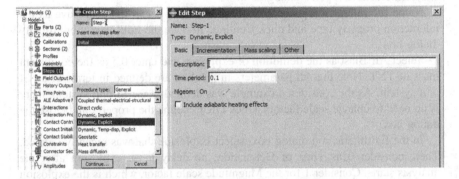

**FIGURE 6.10** Defining step.

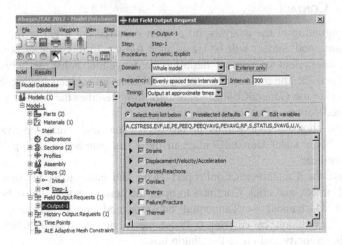

**FIGURE 6.11** Editing field output request.

Select Tools → Datum and define a datum point using Offset from point, and select the vertex, as shown in Figure 6.13. Then consider 0, 0, 0.1 for the offset from the point and click the middle button of the mouse to define a datum.

Select Tools → Reference Point and click the point generated earlier (Figure 6.13).

Double click on Interactions in the ModelTree to open the CreateInteraction dialog box and select IncidentWave and click Continue.

Select the reference point labeled RP-1 as the location of the explosion mass and click CONWEP in the prompt to determine the type of explosion.

Select front surfaces affected by the impact of the explosion by dragging the left mouse button on them and then click the middle mouse button.

After, select one side, the EditInteraction box will be opened. To define the explosion property, click the CreateInteractionProperty icon to open the CreateInteractionProperty dialog box.

Name the interaction property as Explosion and select IncidentWave as the interaction property type and click Continue to open the EditInteractionProperty dialog box.

Select Air Blast as the definition of explosion and enter 0.5 as the equivalent mass of TNT. Note that all parameters in this box are defined in terms of an SI unit system. As the considered example was defined in the SI unit system, there is no need to change scale factors. Click OK to define the property and close the dialog box.

In the EditInteraction dialog box, select Explosion that was recently defined. Then, consider 0 as Time of Detonation; no delay in an explosion after the analysis starts. Consider 1 for the Magnitude scale factor, which is the explosion load generated equal to the real load. Click OK to define the interaction and close the dialog box (Figure 6.14).

### 6.4.10 CONTACT

At the beginning of the analysis, no contact occurred but it led to the solution progressing, and contact will be established further. In this case, it is assumed that the coefficient of friction defining the tangential behavior between the surfaces is 0.35. The most convenient way to define all contacts having the same property is to use GeneralContact. This algorithm identifies the contact areas and the defined contact behavior for all surfaces.

Double click on Interactions in the ModelTree and name the interaction Contact. Then select GeneralContact in the CreateInteraction dialog box and click Continue.

Choose the CreateInteractionProperty icon and name it as Friction. Then, choose Contact, and click Continue to open the corresponding dialog box.

Select Mechanical → Tangential behavior and consider Penalty as the Friction formulation, by entering 0.35 for the friction coefficient. Click OK to define the property and close the dialog box

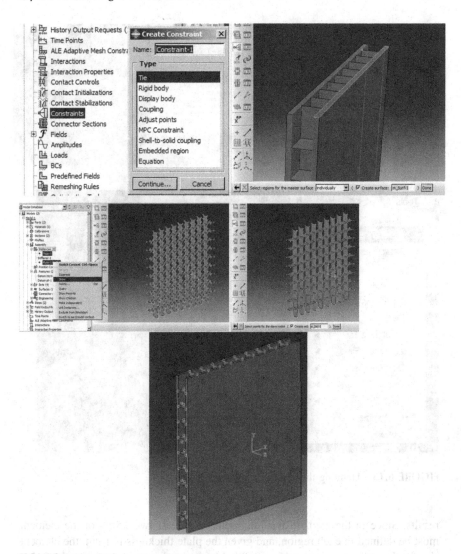

**FIGURE 6.12** Defining tie constraint.

In the EditInteraction dialog box, select Friction, and click OK to apply this definition to all levels (Figure 6.15).

## 6.4.11 MESHING MODULE

At this point, the meshing parts should be developed. In this example, two types of solid and shell geometry are defined, and accordingly, two types of elements must be defined. The explicit method needs a mesh refinement to generate precise

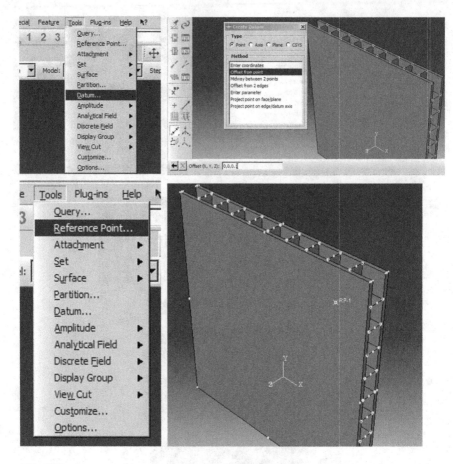

**FIGURE 6.13**  Defining the explosion reference point.

results. Since in the explicit dynamic method, at least two layers of the element must be defined in each region, and given the plate thickness is 1 cm, the element size for the entire model should be 0.005. Moreover, the size is considered suitable for the Stiffener, and the reason why eight elements can model the warping effects produced as a result of the explosion.

Double click on Mesh (Empty) under the Plate in ModelTree to activate the MeshModule. Select Seed → Part to open the GlobalSeeds dialog box and enter 0.005 as the Approximate global size and then click Apply to set the size and click OK to close the dialog box.

Select Mesh → Part and choose Yes in the prompt area to mesh the part (Figure 6.16).

Perform the same procedure for the Stiffener (Figure 6.17).

### 6.4.12 LOAD MODULE

As mentioned earlier, the original sandwich plate model is entirely built-in across all sides. Since only one-quarter of the model is defined, appropriate boundary conditions for the middle surfaces must be considered. The symmetric boundary condition must be used for two intermediate sides, and the clamped boundary condition should be used in the two circumferential sides.

Double click on BCs in the ModelTree to open the CreateBoundaryCondition dialog box. Name the boundary condition XSymmetry and select Symmetry/AntiSymmetry/ENCASTRE as the type and click Continue.

Select the two faces parallel to the YZ plane, as well as the edges in their plane and click the middle mouse button to open the EditBoundaryCondition dialog box. Select XSYMM and click OK to apply and close the dialog box.

Perform the same procedure for a boundary condition named YSymmetry for the two faces parallel to the XZ plane and the edges as YSYMM type. Then, define an ENCASTRE boundary condition named Fixed for four faces and corresponding outer sides (Figure 6.18).

## 6.5 ANALYSIS: JOB MODULE

Finally, a job should be defined to submit the model into the solver.

Double click on the Jobs in the ModelTree to open the CreateJob dialog box. Name it as Explosion and click on Continue to open the EditJob dialog box. Leave the box unchanged and click OK to define the job and close the dialog box. To start the analysis, right click on Explosion under Jobs in the ModelTree and select Submit to begin the analysis (Figure 6.19).

## 6.6 ANALYSIS RESULTS

In this example, deformation of the sheet during and after the explosion should be investigated. Plastic strain contours on the plates and stiffener are also important to predict how they will be deformed. PEEQ, the equivalent plastic strain variable, represents the regions that are scalar and is a good criterion for ductile metal plasticity as defined in Equation (6.3).

$$PEEQ = \sqrt{\frac{2}{3}(\dot{\varepsilon}^{pl} : \dot{\varepsilon}^{pl})} \tag{6.3}$$

where $\dot{\varepsilon}^{pl}$ represents the plastic strain component. This is very important to obtain the variable in the plates and stiffener separately, and why their result ranges differ.

Also, displacement, strain energy, and plastic dissipated energy are very important in describing sandwich plate behavior by the explosion. All results are saved in *.Odb file and can be extracted using the Visualization module.

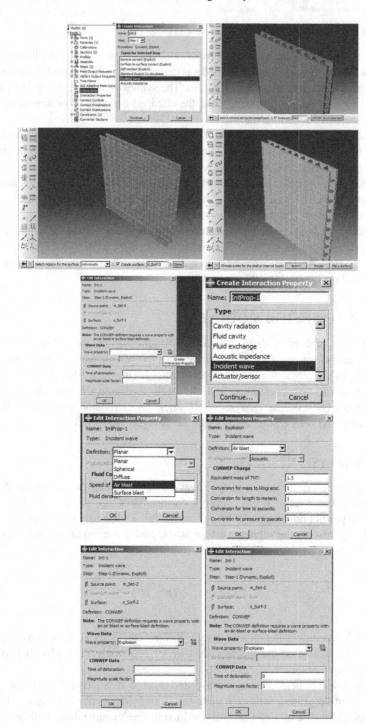

**FIGURE 6.14** Defining explosion interaction.

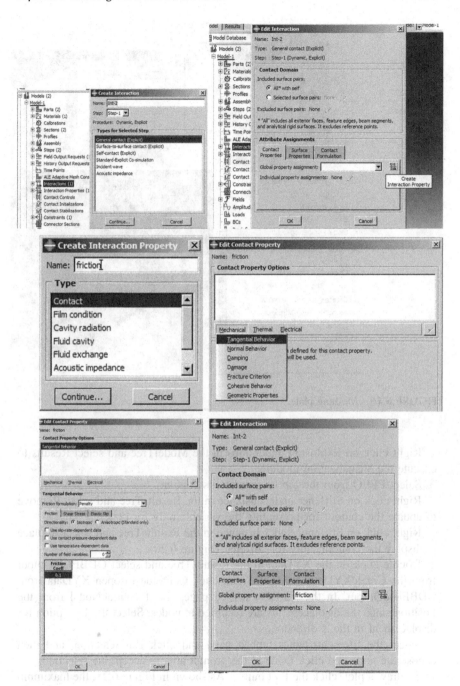

**FIGURE 6.15** Defining contact.

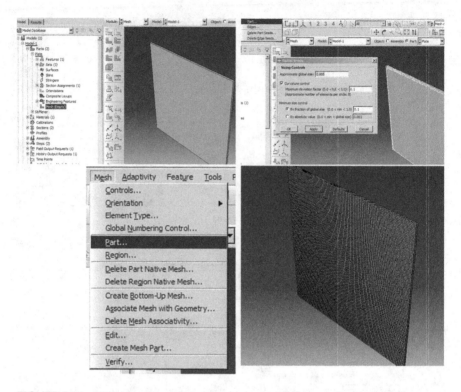

**FIGURE 6.16** Meshing plate.

Right click on Explosion under Jobs in the ModelTree and select Results to execute the Visualization module.

Select PEEQ from the top to generate its contour.

Right click on Stiffener under Instance in the ResultsTree and select Remove to update the contour for presenting the variable only for plates.

Right click on Stiffener under Instance in the ResultsTree and select Replace to display the contour, only for stiffener (Figure 6.20).

Double click on the XYData in the ResultsTree and select ODBFieldOutput from the CreateXYData dialog box and click Continue to open XYData from ODBFieldOutput. In the Variable tabbed page, select UniqueNodal from the Position since the displacements are recorded at nodes. Select the U3 option for displacement in the Z direction.

Select the Elements/Notes tabbed page and click EditSelection, and then choose the node and click Done in the prompt area to accept the selection.

To draw a plot, click the Plot button. As shown in Figure 6.21, the maximum displacement is 7.5 cm.

To extract plastic dissipated energy, double click on the XYData in the ResultsTree and select ODBHistoryOutput from the CreateXYData dialog box

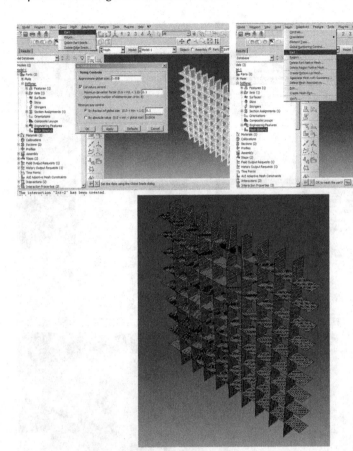

**FIGURE 6.17**   Meshing stiffener.

and click Continue to open XYData from ODBFieldOutput. Select Plastic dissipation as the variable that should be plotted and click Plot. This diagram will display a loss of energy due to plastic behavior. The maximum plastic dissipated energy is estimated to be around 4 kJ.

To extract the strain energy curve, select the same dialog box to select the Strain energy option, and click on Plot. The maximum strain energy increased to 2.7 kJ at first. However, after the damping, it will decrease to about 1 kJ (Figure 6.22).

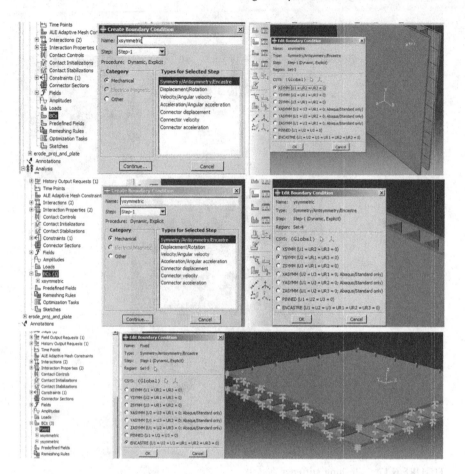

**FIGURE 6.18**  Defining boundary conditions.

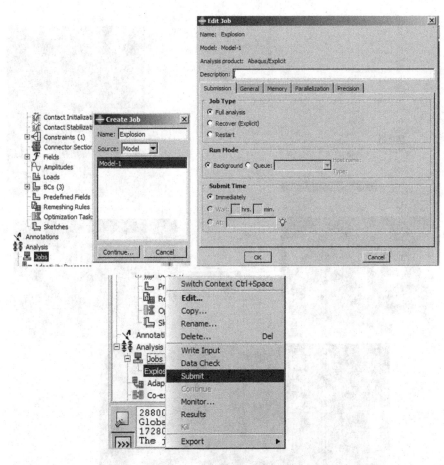

**FIGURE 6.19** Defining the job.

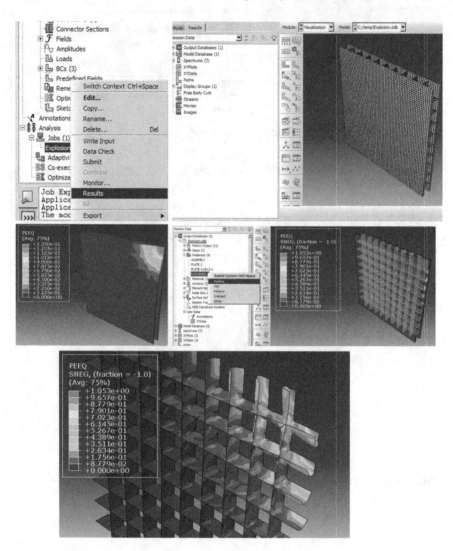

**FIGURE 6.20**  PEEQ contour.

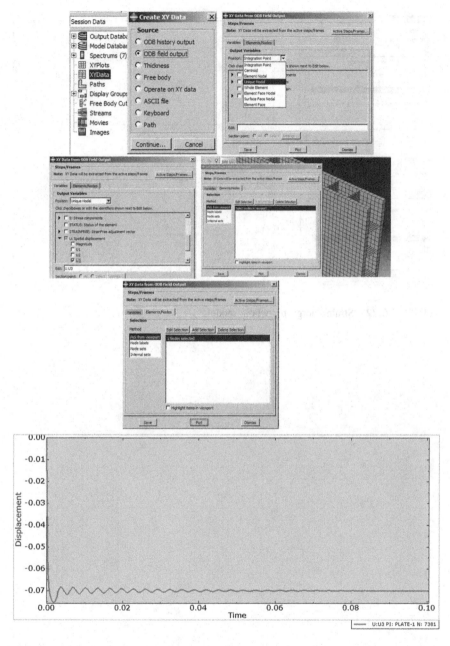

**FIGURE 6.21**  Time history plot of vibration in the sandwich plate center point.

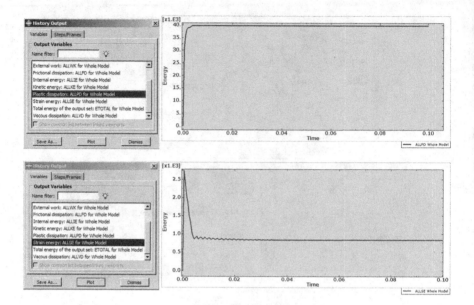

**FIGURE 6.22**   Strain energy for whole model.

# 7 The Impact Load on a Reinforced Concrete Beam

## 7.1 INTRODUCTION

Evaluating structural response under impact loads is considered as one of the most challenging types of structural analysis problems. The impact loads are mostly due to falling of objects on the structural members or accidents.

## 7.2 PROBLEM DESCRIPTION

In this example, the simulation process for a steel-reinforced concrete beam subjected to the impact load is illustrated. The concrete beam is reinforced by longitudinal and transverse bars and located on a roller support at one end and pinned support at another end.

The cross section of the beam is considered to be 300 × 150 mm, which consists of a suitable arrangement of longitudinal and transverse steel bars. A 3-kg impactor falls in the middle of the beam from a height of 10 m (Figure 7.1).

The beam is made of 31 MPa concrete, and all bars are made of steel material, and the yield stress is considered to be 240 Mpa.

## 7.3 OBJECTIVES

1. To simulate RC beam subjected to the impact load
2. To study plasticity and damage occurrence in concrete
3. To investigate the progressive damage to the beam during applied impact load

## 7.4 MODELING

Run Abaqus/CAE from the start menu and close the StartSession dialog box (Figure 7.2).

### 7.4.1 PART MODULE

The geometry should initially be defined. In this problem, the geometry consists of:

DOI: 10.1201/9781003213369-7

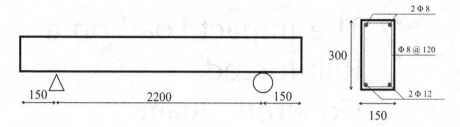

**FIGURE 7.1**   Steel-reinforced concrete beam.

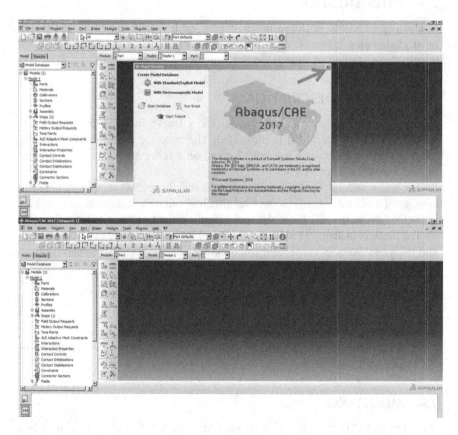

**FIGURE 7.2**   Abaqus/CAE.

- A rectangular beam
  - Two compressive bars with size 8
  - Two tensile bars with size 12

- Twenty-one shear bars with size 8
- A cylinder impactor

These should be drawn as parts. As a 3D problem, all components will be deformed due to the impact of the impactor, which is assumed as a rigid body. In other words, deformations are small compared to the other parts. The beam is considered as solid, and all bars are considered as wire. Their variable gradients are more obvious in longitudinal rather than transverse, and this is why their lengths are larger than their widths. This means that the problem is cost-effective and faster to run, with good precision. Cylinder impactor is considered as a rigid shell.

### 7.4.1.1  Beam

Double click on Parts in the ModelTree to open CreatePart dialog box. Name the part as Beam and set Approximate size 1. Leave other defaults unchanged and click Continue to activate Sketcher.

Select Create Lines: [Rectangle] and draw a rectangle. Select AddDimension and determine the plate dimensions as a 0.3 × 0.15 square.

Click the middle mouse button twice to open EditBaseExtrusion and enter 2.5 as the Depth. Click OK to apply and close the dialog box (Figure 7.3).

### 7.4.1.2  Longitudinal Rebar

Double click on Parts in the ModelTree to open the CreatePart dialog box and name the part RebarTop. Choose Wire shape and set the Approximate size to 5. Then click on Continue to activate the sketcher.

Select Create Lines; Connected and draw a horizontal line.

Select AddDimension and set the length of the line to 2.5. Click the middle mouse button twice to extract the line (Figure 7.4).

To define the lower rebar, there is no need to redraw them, because the upper and lower bars are only different in terms of cross section, and are of the same length.

Right-click on RebarTop under Parts in the ModelTree and select Copy, and name it RebarBottom, and then click OK (Figure 7.5).

### 7.4.1.3  Shear Bar

For creating a Shear bar, it should be noted that the transverse concrete cover is 3 cm from each side of the shear bar. Also, for easier modeling, the arches were ignored on the corners, and an angle of 90 degrees will be plotted. This does not affect the solutions.

Double click on Parts in the ModelTree to open the CreatePart dialog box and name the part Shear bar. Choose Wire shape and set the approximate size to 5. Then click on Continue to activate the sketcher.

Select Create Lines: [Rectangle] and draw a rectangle.

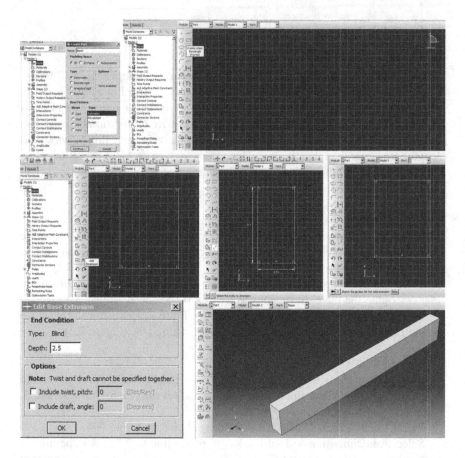

**FIGURE 7.3**   Creating the beam.

Select AddDimension and set the height and width as 0.24 and 0.09, respectively. Click the middle mouse button twice to extract the rectangle (Figure 7.6).

### 7.4.1.4   Impactor

The impactor is a stiffened cylinder that falls on the beam. With regard to the stiffness of the impactor, in comparison to the other parts, its deformations have been neglected. In other words, the concept of a rigid body can be used to define the impactor.

A rigid body is a set of nodes and elements whose movement is controlled by the movement of a single node which is called the rigid body reference node.

The rigid body movement can be determined by applying boundary conditions in the reference node of the rigid body. Loading on a rigid body is obtained from the load applied to the rigid body reference node.

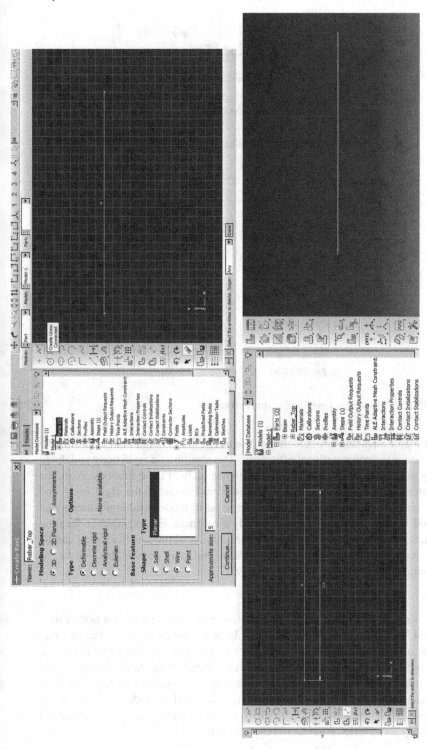

**FIGURE 7.4** Creating an upper longitudinal rebar.

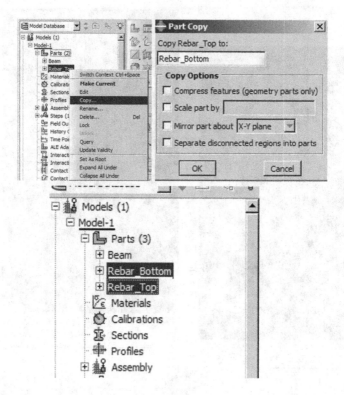

**FIGURE 7.5**  Creating a lower longitudinal rebar.

The major advantage of models that use a rigid body rather than a deformable finite element is computational efficiency. The motion of a rigid body is defined only by six degrees of freedom on the reference body of the rigid body. Also, the concentrated loading is located on the reference node of the rigid body and loaded on rigid elements. Rigid bodies are used to model hard components that are fixed in motion or under the rigid body movement. For example, in forming processes, modeling tools (such as punches, molds, clamps, rollers, etc.) can be considered as rigid in solving the problem and may reduce the volume of computations.

### 7.4.1.5  Components of a Rigid Body

The movement of a rigid body is controlled by the movement of a node called the rigid body reference node. The rigid body reference node has translational degrees of freedom and rotational degrees of freedom and must be defined for each rigid body separately. In each case, the reference node should be considered where it is placed on the proper axis of the object. In addition to the rigid body reference node, the *discrete rigid body* is composed of nodes formed by the allocation of elements and nodes to the rigid body. These nodes which are called RigidBodySlaveNodes connect to other elements.

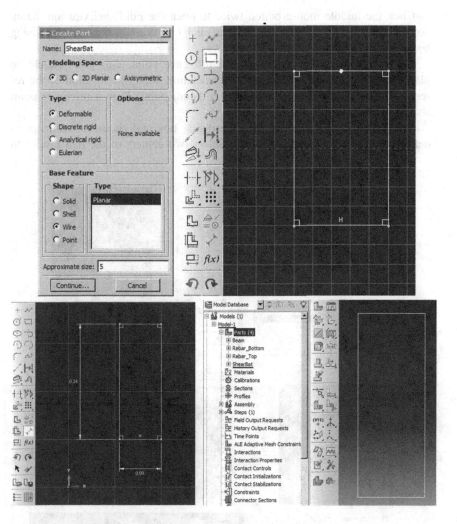

**FIGURE 7.6**   Creating a shear bar.

In this example, a cylindrical impactor is assumed as a half-cylindrical discrete rigid body, that should be considered as the Shell, and a reference point should be defined on its center.

Double click on Parts in the ModelTree to open the CreatePart dialog box and name the part Impactor. It should be Discrete rigid, a Shell shape, and Extrusion type. Set the approximate size to 5, and then click on Continue to activate the Sketcher.

Select Create arc; Center and twoEndpoints and draw a half-circle.

Select AddDimension and set the half-circle radius to 0.1.

Click the middle mouse button twice to open the EditBaseExtrusion dialog box and enter 0.15 for the Depth and then click OK to apply and close the dialog box (Figure 7.7).

For a rigid body, it needs a defined reference point. This point should be placed in the center of the impactor's center line. First, the location of the reference point should be determined by a datum point. Then a reference point should define it.

Select Tools → Datum to open the dialog box and choose a datum point defined Midway between two points. Select two centers of the half circles to

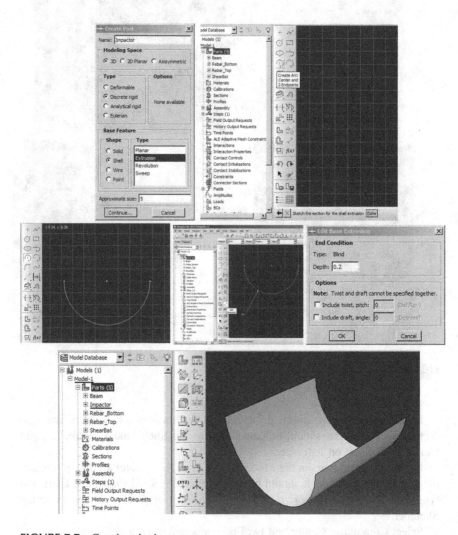

**FIGURE 7.7**  Creating the impactor.

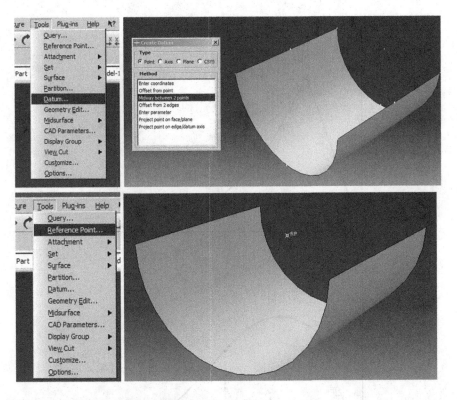

**FIGURE 7.8** Defining the rigid body reference node.

determine the midpoint between two points. Select Tools → ReferencePoint and click on Datum (Figure 7.8).

### 7.4.2 MATERIAL PROPERTIES

In this example, two materials are used: Concrete with a compressive strength of 31 MPa and steel with a yield stress of 240 MPa. Therefore, there are two different material models. Concrete is a brittle material in which the compressive and tensile behaviors are different, while steel is a ductile material, and its compressive and tensile behaviors are assumed to be the same.

Abaqus software introduces different material models for defining concrete properties. The most accurate material model for the definition of concrete is ConcreteDamagedPlasticity where it will be entered in the Plastic flow along with the behavior of pressure plastic, tensile plastic behavior, and elastic behavior. Also, this model provides the ability to add tensile and compressive damage separately. To define this model, concrete compression and tensile behavior should be determined from the laboratory test results (or based on equations).

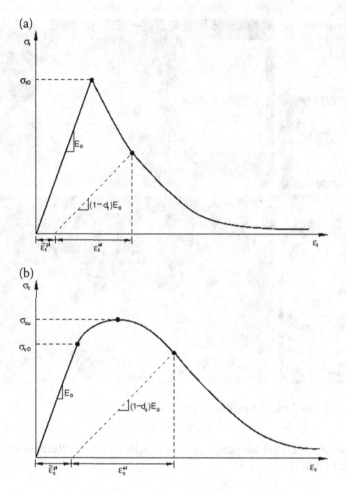

**FIGURE 7.9** (a) Tensile and (b) compressive behavior of concrete.

The test results show the pressure and tensile test for concrete single-axle load (Figure 7.9).

The parameters shown in Table 7.1 represent the concrete parameters in the ConcreteDamagedPlasticity model.

In this example, the material properties are extracted from the results of the compressive and tensile test with a compressive strength of 31 MPa. The results consist of elasticity properties, tensile behavior, compressive behavior, and plastic flow represented in Tables 7.2, 7.3, and 7.4.

Then, mechanical properties for steel should be defined from the tensile test (Tables 7.5 and 7.6).

**TABLE 7.1**

**Material constants for concrete 31 MPa**

| Parameter | Sign |
|---|---|
| Tensile stress | $\sigma_t$ |
| Tensile yield stress | $\sigma_{t0}$ |
| Young's modulus | $E_0$ |
| Damage parameter | $d_t$ |
| Tensile equivalent plastic strain | $\bar{\varepsilon}_t^{pl}$ |
| Tensile elastic strain | $\varepsilon_t^{el}$ |
| Compressive stress | $\sigma_c$ |
| Compressive yield stress | $\sigma_{c0}$ |
| Ultimate compressive stress | $\sigma_{cu}$ |
| Compressive equivalent plastic strain | $\bar{\varepsilon}_c^{pl}$ |
| Compressive elastic strain | $\varepsilon_c^{el}$ |

**TABLE 7.2**

**Elasticity properties for concrete 31 MPa**

| Young's modulus (GPa) | Poisson's ratio | Density $\left(\frac{kg}{m^3}\right)$ |
|---|---|---|
| 25 | 0.2 | 2,400 |

**TABLE 7.3**

**Plasticity properties for concrete 31 MPa**

| Compressive behavior | | | Tensile behavior | |
|---|---|---|---|---|
| Yield stress (Pa) | Plastic strain | Damage parameter | Yield stress (Pa) | Plastic strain |
| 9,300,000 | 0 | 0 | 3,507,692 | 0 |
| 23,003,533 | 9.38E-05 | 0.030122 | 2,336,335 | 0.00029 |
| 31,000,000 | 0.000752 | 0.155963 | 1,917,857 | 0.00055 |
| 15,791,944 | 0.003394 | 0.620767 | 1,681,418 | 0.000802 |
| 9,632,989 | 0.004959 | 0.796765 | 1,523,177 | 0.001052 |

## TABLE 7.4
### Plastic flow properties for concrete 31 MPa

| Viscosity parameter | K | fb0/fc0 | Eccentricity | Dilation angle |
|---|---|---|---|---|
| 0.001 | 0.667 | 1.16 | 0.1 | 32 |

## TABLE 7.5
### Elasticity properties for steel

| Young's modulus (GPa) | Poisson's ratio | Density $\left(\frac{kg}{m^3}\right)$ |
|---|---|---|
| 200 | 0.3 | 7,800 |

## TABLE 7.6
### Plasticity properties for steel

| Yield stress (Pa) | Plastic strain |
|---|---|
| 425,000,000 | 0 |
| 600,000,000 | 0.21 |

Double click on Materials in the ModelTree and name it Concrete. Then enter the material data according to Figure 7.10. Finally, click OK to define the material.

Again, double-click on Materials in the ModelTree and name it as Steel. Then enter the material data according to Figure 7.11. Finally, click OK to define the material.

### 7.4.3 SECTION PROPERTIES

It is necessary to introduce a section for each property. These sections include a solid section for the beam, a beam section for the shear bar, and two truss sections for the longitudinal bars.

Double click on Sections in the ModelTree and open the CreateSection dialog box. Name the section as Concrete_Beam and consider a Solid and Homogenous type. Click Continue to open the EditSection dialog box and in the EditSection

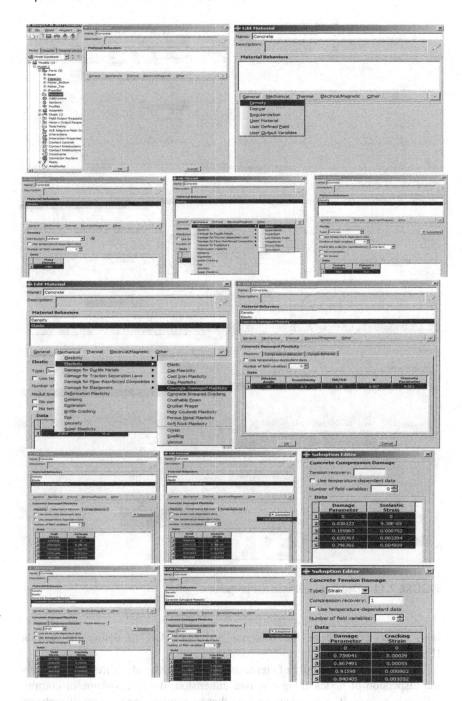

**FIGURE 7.10** Defining concrete material properties.

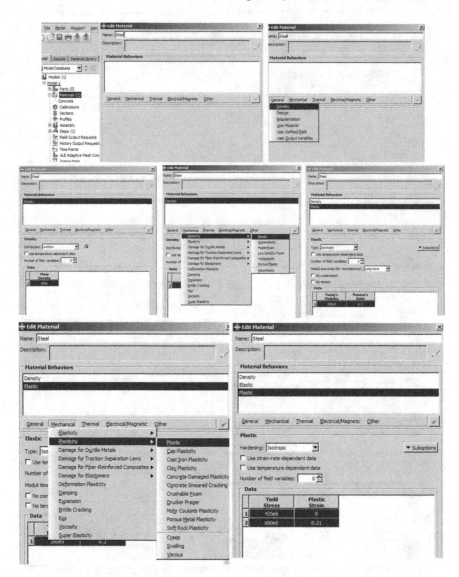

**FIGURE 7.11** Defining steel material properties.

box, select Concrete as the Material and click OK to define the section (Figure 7.12).

Abaqus presents a family of elements under the name of a wire that considers the dispersion of nodes along just one dimension, thus the volume of computations is reduce. It is necessary to note that by minimizing the size of matrices and the degree of freedom of the model resulting from the simplification of the model, the assumption of making a more precise model and reviewing the model

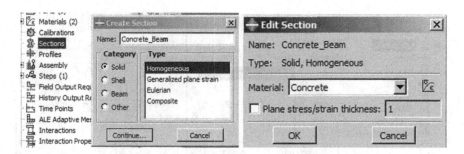

**FIGURE 7.12** Defining a solid section for a concrete beam.

completely in these situations should be considered. If these assumptions are not addressed, the response may be uncertain.

Wire elements are only used in structural modeling, where the dimension of one direction is considerably longer compared to the other two. It is possible to ignore variations of the variables on the cross section of the elements. The criterion for using beam elements in component modeling is that for ductile metals the length of a geometry dimension should be more than 10 times the length of another dimensions. However, with shell elements, this ratio should be used with caution.

Therefore, in this example, to simulate longitudinal and transverse bars, wire elements can be used and, therefore, a smaller volume computations will be made. Transverse bars will be considered as the Beam, and longitudinal bars will be the Truss.

Double click the Sections option and name it ShearBar. Then select Beam as the Category and Beam as the Type. This section provides the ability to define flexural, torsional, and shear behavior, along with compressive and tensile stress variations. Then click Continue to open the EditSection dialog box. In the dialog box, select Steel as the Material name and click on the CreateBeamProfile icon to open CreateProfile dialog box to define the cross section of the shear bar. Choose Circular as the Shape and click Continue to open EditProfile. Enter the bar radius as 0.004 and click OK to define the profile. Then, click OK in the EditBeamSection box to create the section (Figure 7.13).

Double click on the Sections and name it as Rebar_Top and consider Beam as the Category and Truss as the Type and click Continue to open the EditSection dialog box. Select Steel as the Material name and as shown in Figure 7.14, enter 5e-5 as the Cross-sectional area as given in Equation (7.1).

$$A = \pi (0.004^2) = 5e - 5 \tag{7.1}$$

Perform the same procedure as Rebar_Bottom for the lower longitudinal bars and enter 1.13e-4 as its Cross-sectional area (Figure 7.15).

For the impactor, due to the rigidity assumption, it is not necessary to define the section properties.

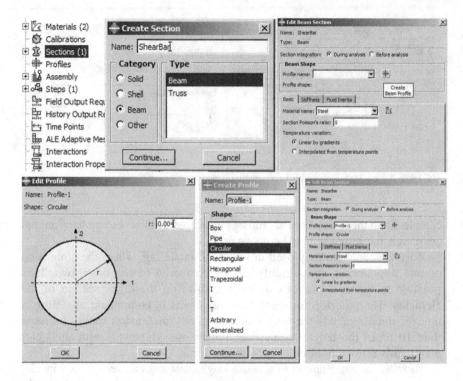

**FIGURE 7.13** Defining a beam section for the shear bar.

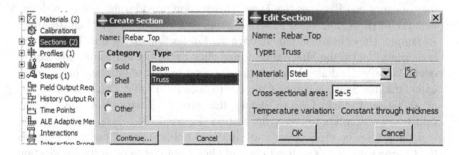

**FIGURE 7.14** Defining a beam truss section for the top rebar.

### 7.4.4 SECTION ASSIGNMENT

In the next step, the defined sections should be assigned to their corresponding parts.

Double click on SectionAssignments under Beam in the ModelTree, after selecting all geometry and click Done in the prompt area. Then select

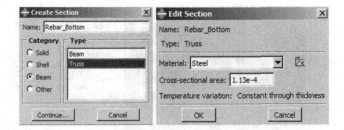

**FIGURE 7.15**  Defining a beam truss section for the top rebar.

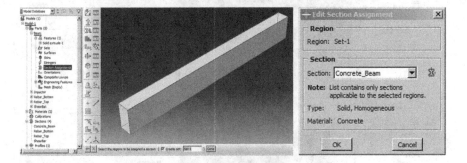

**FIGURE 7.16**  Assigning Concrete_Beam section.

Concrete_Beam in the EditSectionAssignments dialog box and select OK to assign the section (Figure 7.16).

Perform the same for other parts except for Impactor (Figure 7.17).

It is necessary to note that regarding the nondeformation of Impactor, there is no need to define and assign a section.

### 7.4.4.1 Assigning Beam Normal

Given that the beam type determines the section of the transverse bars, it is necessary to determine the norm of the beam to correctly calculate the moment of inertia and the cross-sectional orientation.

To determine the angle and position of the beam profile along its length on each link, the Abaqus requires the definition of a three-axis coordinate system consisting of two perpendicular vectors to the beam inside the section, as well as a tangent vector. The two normal vectors perpendicular to the beam are referred to as n1 and n2 local vectors, known as beam normals, and tangential vector on the beam with t, which follow the law of the right hand, as shown in Figure 7.18.

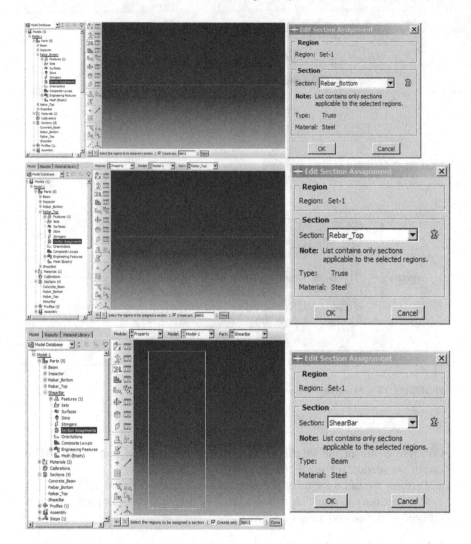

**FIGURE 7.17**  Assigning section.

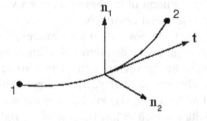

**FIGURE 7.18**  n1, n2, and the vectors that determine the position of the profile on a beam.

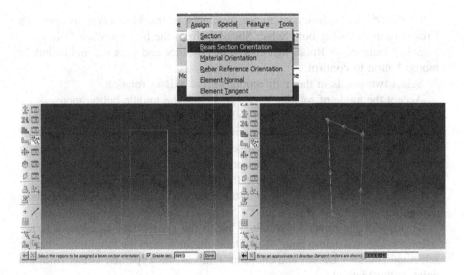

**FIGURE 7.19** Determining n1 direction for the shear bar.

The user determines the direction for the vector n1. Determining the direction of the tangent vector t refers to the geometry, and the n2 vector is obtained by using the external multiplication of the two vectors in accordance with Equation (7.2).

$$n2 = t \times n1 \qquad (7.2)$$

It is not easy to determine the beam normal in some applications. Depending on the geometric condition of the problem, there are various methods to determine it, but in this example, given the exact coordinates are available, this direction can easily be introduced.

Select the Assign → BeamSectionOrientation option and then select the Shear bar entirely by dragging it and click the middle mouse button. Accept the default direction (0, 0, −1) as the orientation of the n1 vector and again click the middle mouse button to confirm. The red arrows shown in Figure 7.19 show the direction of the normal and tangent vectors. Click OK on the prompt area to confirm.

### 7.4.5 Assembly Module

It is necessary to assemble the model and combine all the parts together correctly. This is needed to create instances in the Assembly and to place them at the correct place. For convenience, put the steel parts together first and, then, the concrete beam to cover them.

Double click the Instances under Assembly in the ModelTree to open the CreateInstances dialog box. Select ShearBar from the box and click OK.

Select Instance → Rotate and choose the instance and click the middle button mouse button to confirm.

Select two points in the Y direction, for the axis of rotation.

Accept the angle of rotation as 90° and click the middle button mouse.

Select Instance → LinearPattern and choose Shear bar. Then click the middle mouse button to open the LinearPattern dialog box. In Direction 1 box, consider 21 and 0.12 as the Number and Offset, respectively. Also, set the Number in Direction 2 to 1, where it will then produce a pattern in the direction disable. Finally, click OK to apply and close the dialog box (Figure 7.20).

Continue to add the Rebar_Top instance by selecting it in the CreateInstance dialog box.

To transfer the upper longitudinal bar to the appropriate location, it is necessary to create a datum point at a suitable distance from the longitudinal bar by selecting Tools → Datum and choosing Point as the Type and Offset from the point as the Method.

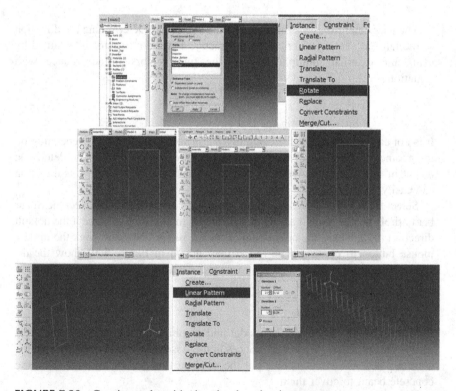

**FIGURE 7.20** Creating and positioning the shear bar instance.

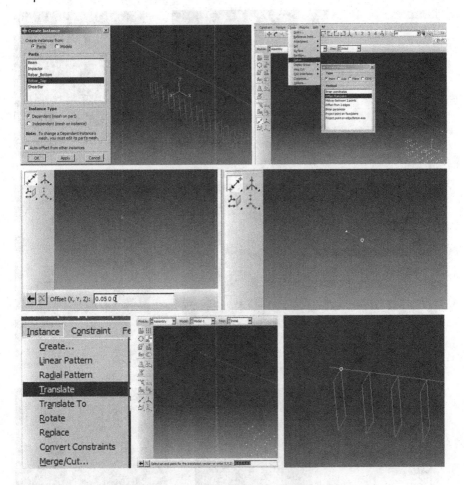

**FIGURE 7.21** Creating and positioning the upper longitudinal bar instances.

Select the end of the Rebar_Top and click the middle mouse button and enter 0.05, 0, 0 as the offset and click the middle mouse button again.

Select Instance → Translate and choose the bar and click the middle mouse button. Then select the datum point as the point of the beginning of translate, and select one of the upper corners of the first shear bar to fix the rebar to fit in place.

Perform the same for another Rebar_Top (Figure 7.21).

Perform the same for the two lower longitudinal bars (Figure 7.22).

From the CreateInstances dialog box, click on the Beam and then OK to add it to the assembly.

Rotate the beam by 90° around one of the vertical axes by selecting two points in its direction.

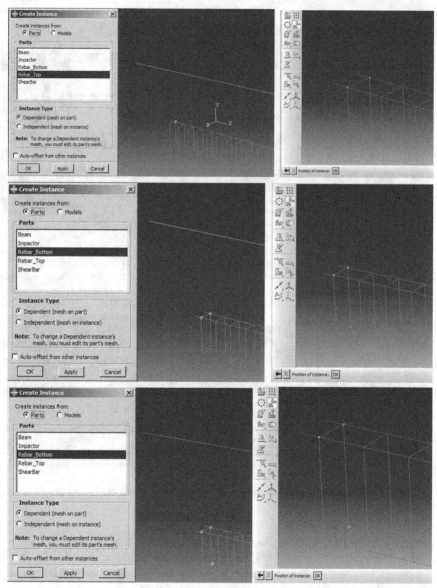

**FIGURE 7.22** Creating and positioning the longitudinal bar instances.

Create a datum point using Offset from point method, and the upper corner of the beam as 0, −0.03, 0.03.

Translate the beam from the datum point created to the appropriate vertex on the bar (Figure 7.23).

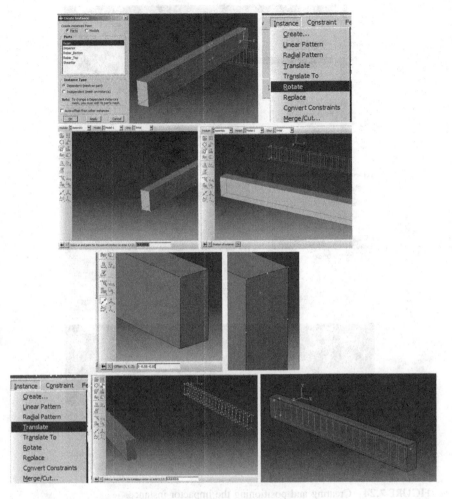

**FIGURE 7.23**  Creating and positioning the beam instance.

Select the Impactor from the CreateInstances dialog box and click OK to add it to the assembly.

Translate the impactor from the lower point to the middle of the upper edge of the beam. The final assembly form is obtained (Figure 7.24).

### 7.4.6  Step Module

At this step, it is necessary to define the type of resolution.

This case is an example of dynamic analysis because many variations in the status of the model are created in such a short time. Abaqus software can carry out dynamic solutions in two ways.

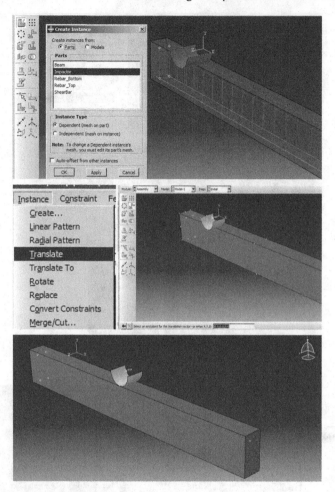

**FIGURE 7.24**   Creating and positioning the impactor instance.

### 7.4.6.1 Implicit Dynamics

Using this method, the response in each load increment is calculated based on the matrix method, and depending on the number of degrees of freedom, the required interfaces, etc., the computational volume is determined.

Therefore, given the approach of the implicit solution method for solving equations simultaneously, made this technique as a powerful method for analyzing a wide range of issues.

### 7.4.6.2 Explicit Dynamic

In this method, as used in the Abaqus/Explicit solver software, the equations are solved sequentially based on the central difference process.

Double click on the Steps in the ModelTree and open the CreateStep dialog box. Select Dynamic, Explicit from the box and click Continue to open the EditStep dialog box.

Set the Time period to 0.2 seconds and click OK to close the dialog box (Figure 7.25).

### 7.4.7  Request Output

All outputs requested by Abaqus are appropriate for the analysis, but compressive damage should be added to its list. Also, the output is preferred to be recorded using many time intervals to produce results.

Double click on the F-Output-1 under FiledOutputRequests in the ModelTree to open the EditFieldOutputRequest and enable DAMAGEC from the list of variables. Then, change the interval of outputs from 20 to 200. This will result in more quality charts (Figure 7.26).

### 7.4.8  Interaction Module

In this example, two types of constraints should be used. The first is the EmbeddedRegion type to embed the steel bars by concrete. To define the constraint, first, concrete should be hidden, and the second is the Coupling type to couple some nodes of surface/surfaces to a control point.

To define the embedded region constraint, the beam should be hidden at the beginning. Then, all steel bars should be selected as an embedded and redisplay beam, and select as the host region.

To define the coupling constraint in the supports, areas corresponding to the boundary conditions should be defined in the PartModule. This is performed by partitioning in this module. Then midpoints of each of these surfaces are selected as their control point. Finally, the boundary conditions of the hinge and the roller are determined on these points.

### 7.4.8.1  Embedded Region

Right click on the Beam under Instances in the ModelTree and select Hide.

Double click on Constraints in the ModelTree to open the CreateConstraint dialog box and select EmbeddedRegion as the Type. Then click Continue.

Select all the steel bars by dragging them as parts that are embedded in concrete and click the middle mouse button to confirm the selection.

Select the SelectRegion option from the prompt area as the type of host region.

Right click on Beam under Instances in the ModelTree and select Show to redisplay it.

Select Beam as the host region part and click the middle mouse button to open the EditConstraint dialog box. Accept all defaults and click OK to define the constraint (Figure 7.27).

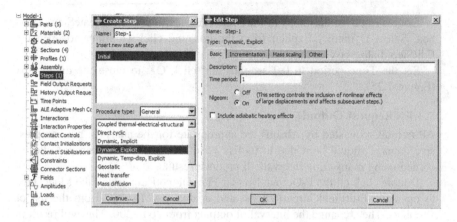

**FIGURE 7.25**   Defining the analysis step.

### 7.4.8.2   Coupling

Double click on Beam under Parts in the ModelTree to make it current. Select Tools → Partition and choose Face as the Type and Sketch as the Method. Then rotate the beam so that the lower face is displayed. Select the face as a sketcher face, and one of the end edges as the sketch axis to open sketcher.

Draw two rectangles by Create Lines: Rectangle on the face.

Select AddDimension and determine the width of the rectangles as 0.1 and the distance to the end edges as 0.1.

Click the middle mouse button twice and exit sketcher (Figure 7.28).

Select Tools → Datum to define a datum point midway between the two points Method. Click the points in the middle of the edges that are created by the partitions.

Perform the same for another surface (Figure 7.29).

Select Interaction as the current Module.

Select Tools → ReferencePoint and choose both datum points created recently (Figure 7.30).

Double click on Constraints in the ModelTree and select Coupling in the CreateConstraint dialog box. Then click Continue.

Select one of the reference points as a control point and click the middle mouse button.

Select the Surface as the surface type in the prompt area and click on the corresponding surface. Then click the middle mouse button to open the EditConstraint dialog box. Accept all setting and click OK to define the coupling constraint.

Perform the same for another reference point (Figure 7.31).

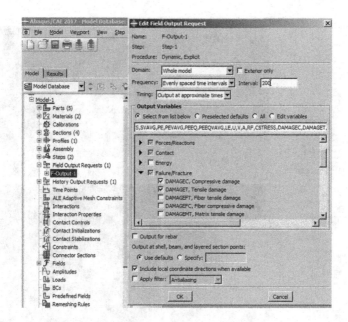

**FIGURE 7.26**   Editing field output.

### 7.4.8.3 Contacts

Contact will be established when nodes of a surface penetrate the areas between the nodes of another surface. The first surface should be stiffened material rather than the second.

In this example, the contact between the impactor and the concrete beam is evident. However, the slip or contact between the internal sections of the beam is almost zero and can be ignored (the embedded region constraint removes the possibility of slipping between the reinforcement bars and concrete). Contact between the impactor surface and the upper surface of the beam with a variable friction coefficient in terms of slip rate and friction coefficients in stationary and moving states is given in Equation (7.4).

$$\mu = \mu_k + (\mu_s - \mu_k)e^{-d_c \dot{\eta}_{eq}} \tag{7.4}$$

Contact will be established when nodes of a surface penetrate the areas between the nodes of another surface. The first surface should be stiffened material rather than the second.

In this example, the contact between the impactor and the concrete beam is evident. However, the slip or contact between the internal sections of the beam is almost zero and can be ignored (the embedded region constraint removes the possibility of slipping between the reinforcement bars and concrete). Contact between the impactor surface and the upper surface of the beam with a variable

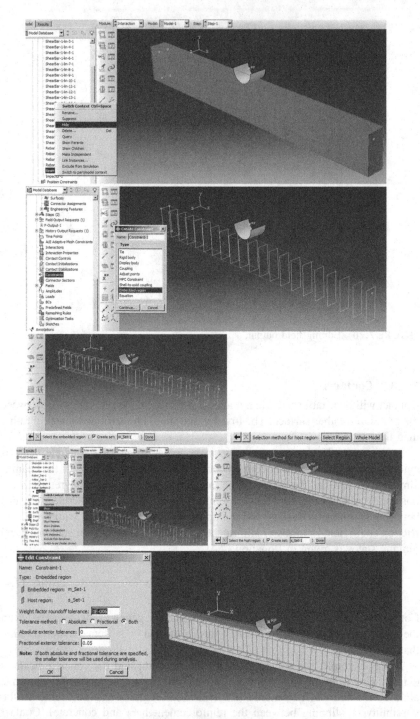

**FIGURE 7.27** Defining the embedded region constraint.

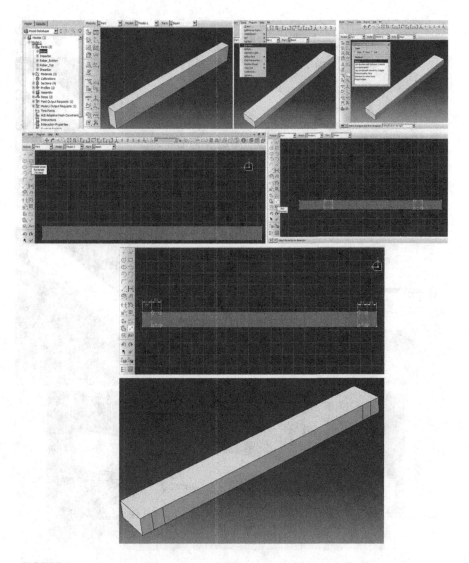

**FIGURE 7.28** Beam partitioning to make boundary conditions surfaces.

friction coefficient in terms of slip rate and friction coefficients in stationary and moving states is given in Equation (7.5).

$$\mu = \mu_k + (\mu_s - \mu_k)e^{-d_c\dot{\eta}_{eq}} \qquad (7.5)$$

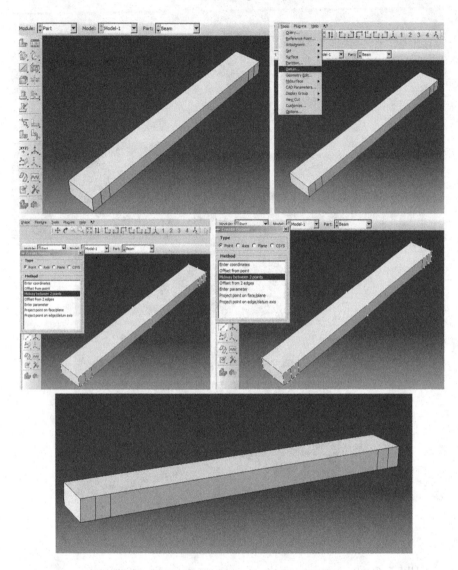

**FIGURE 7.29** Defining the datum point in the middle of the surfaces created by partitioning.

where $\mu$ is the equivalent coefficient of friction. $\mu_k$ is the friction coefficient of motion. $\mu_s$ is the static friction coefficient. $d_c$ is the coefficient of descent. $\dot{\eta}_{eq}$ is the slip rate of the surfaces at each moment of analysis.

In the example, the surface-to-surface contact will be applied between the impactor and the upper face of the beam so that the outer surface is considered master, given the rigidity and upper beam face are considered slave.

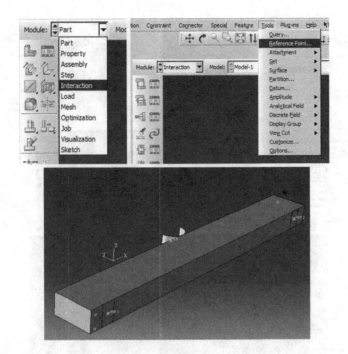

**FIGURE 7.30** Converting datum points to reference points.

Double click on the Interaction in the ModelTree to open the CreateInteraction dialog box. Select Surface-to-Surface as a contact type, and then click Continue.

Select impactor as the first surface and click the middle mouse button. Then from the prompt area, choose a color that is related to the outer side (that is, Brown).

Select Surface as the type of the second surface and choose the upper face of the beam and click the middle mouse button to open the Edit Interaction dialog box.

Click the Create Interaction Property icon to open the dialog box and choose Contact as the type and click Continue to open the Edit Contact Property dialog box.

Select Mechanical → Tangential behavior and choose Static-Kinetic Exponential Decay as Friction formulation. Enter the formulation parameters, as shown in Table 7.7, and then click OK to close the dialog box.

Click OK to close EditInteractionProperty dialog box (Figure 7.32).

### 7.4.8.4 Point Mass

As this example is a dynamic analysis, the mass of all components must be determined. However, due to the definition of density of all of the variable parts,

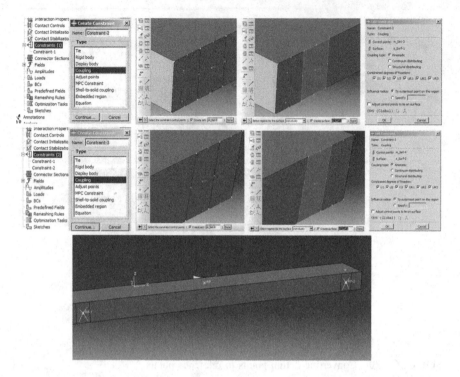

**FIGURE 7.31** Defining the coupling constraint.

it is not necessary to define an additional item for them, except in the impactor. Due to the assumption of rigidity and the impossibility of defining the properties of materials for it, the mass must be directly determined for the impactor.

In this example, the mass of the impactor is 3 kg, and the fall height is 5 m.

Select Special → Inertia → Create to open the CreateInertia dialog box. Choose PointMass/Inertia and select the reference point of the impactor. Then click the middle mouse button to open the EditInertia dialog box. Enter 3 for isotropic mass and click OK to apply and close the dialog box (Figure 7.33).

### 7.4.9   Load Module

The beam supports are considered to be roller and hinged. The impactor should be controlled to lateral displacement and constraint in translating just in the Y direction and with no rotations.

### 7.4.9.1  Hinge Boundary Condition

Double click on the BCs in the ModelTree to open the CreateBoundaryCondition. Consider Displacement/Rotation as the type, and then click Continue.

**TABLE 7.7**

**Parameters of static-kinetic exponential decay formulation**

| $\mu_s$ | $\mu_k$ | $d_c$ |
|---------|---------|-------|
| 0.4     | 0.2     | 4     |

Select the reference point on the left side of the beam and click the middle mouse button to open the EditBoundaryCondition box. Disable all degrees of freedom by checking them except UR3, and then click OK (Figure 7.34).

### 7.4.9.2 Roller Boundary Condition

Define the second boundary condition as a Displacement/Rotation type for another reference point and disable all degrees of freedoms by checking all components except U1 and UR3 (Figure 7.35).

Perform the same for displacement/rotation boundary condition for the reference point of the impactor and restrain all degrees of freedom except U2 (Figure 7.36).

### 7.4.10  Predefined Field

It is assumed that the impactor has fallen from a height of 10 m. The impactor's velocity at the moment of fall is given by (7.6).

$$V = 2\sqrt{5 \times 9.81} \cong 14\text{m/s} \tag{7.6}$$

Double click the PredefinedFields in the ModelTree to open the CreatePredefinedField dialog box and select Velocity as the Type and click Continue.

Choose the impactor's reference point and click the middle mouse button to confirm and open the EditPredefinedField.

Consider −14 for V2 and click OK to apply and close the dialog box (Figure 7.37).

### 7.4.11  Mesh Module

In finite element models, make sure that the meshing size is the same (or approximate) in all sectors. In the model, four types of elements are used as:

- The solid element for Beam
- The rigid element for Impactor
- Truss element for longitudinal bars
- Beam element Shear bar

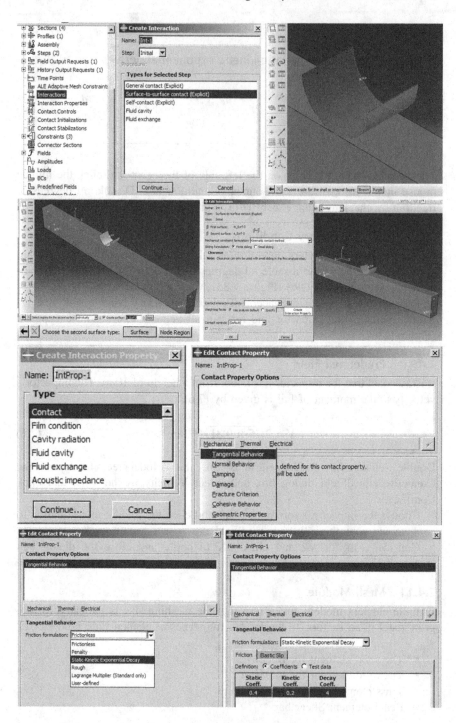

**FIGURE 7.32** Defining contact.

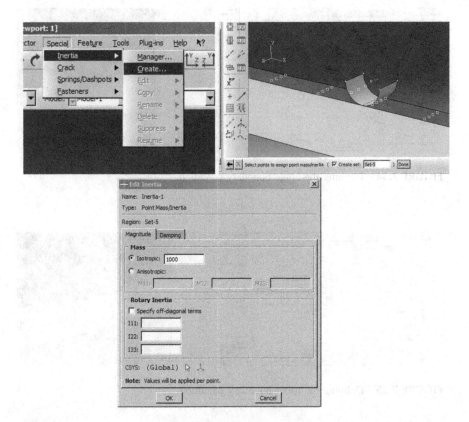

**FIGURE 7.33** Defining point mass for impactor.

However, the element size is considered the same. Note that all the elements used are defined in the Abaqus/Explicit solver library.

### 7.4.11.1 Beam

Double click on the Mesh (Empty) under Beam to activate the MeshModule and then select Seed → Part to open GlobalSeeds. Set the ApproximateGlobalSize as 0.02 and click Apply to display nodes almost positions.

Select the Mesh → Part and click Yes in the prompt area to mesh the part by using 15,000 elements (Figure 7.38).

### 7.4.11.2 Impactor

Double click on Mesh (Empty) under Impactor in the ModelTree to activate the MeshModule, and set the approximate size to 0.02 and mesh the part (Figure 7.39).

Double click on the Mesh (Empty) under Rebar_Botom and do the same for the Beam, set the size to 0.02.

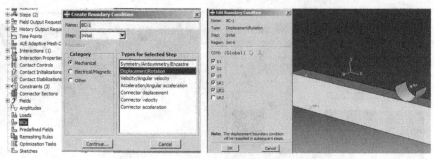

**FIGURE 7.34**   Defining the hinge boundary condition.

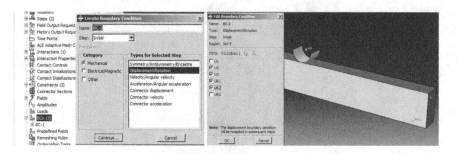

**FIGURE 7.35**   Defining the roller boundary condition.

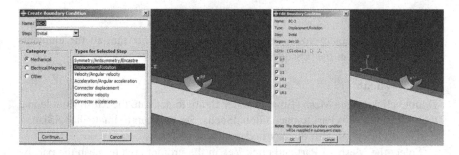

**FIGURE 7.36**   Defining the impactor boundary condition.

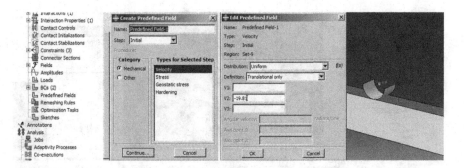

**FIGURE 7.37**    Defining the predefined field for impactor's velocity.

Select Mesh → ElementType and select the part and click the middle mouse button to open the ElementType dialog box. Choose Truss as the Family and as the T3D2 element type. Click OK to apply and close the dialog box (Figure 7.40).

Double click on Mesh (Empty) under Rebar_Top and set the mesh size as the same as the other components, that is 0.02, and consider the T3D2 element for it as the Type (Figure 7.41).

Double click on Mesh (Empty) under ShearBar and set the mesh size the same as the other components, that is 0.02, and consider B31 as the element type in the BeamFamily element for it (Figure 7.42).

## 7.5  ANALYSIS: JOB MODULE

In the next step, a job must be defined to solve the problem and then submitted to obtain the results.

Double click on Jobs in the ModelTree to open the CreateJob dialog box and name it as Impact and then click Continue to open the EditJob dialog box. Click OK to accept all the settings and close the dialog box.

Right click on the job and select Submit to begin the analysis (Figure 7.43).

## 7.6  ANALYSIS RESULTS

In this example, the stresses and damage on concrete and stress on longitudinal bars, and the displacement time history at the beam's middle point and the contour of the equivalent plastic strain have been requested. Note that all contours included the results at the end of the analysis (at $t = 0.2$ seconds).

Right click on the Impact (Completed) under Jobs in the ModelTree and click on Results to activate the Visualization module.

To extract the stress values on the concrete, right click the Beam under Instances in the ResultsTree and select the Replace option to display only the part.

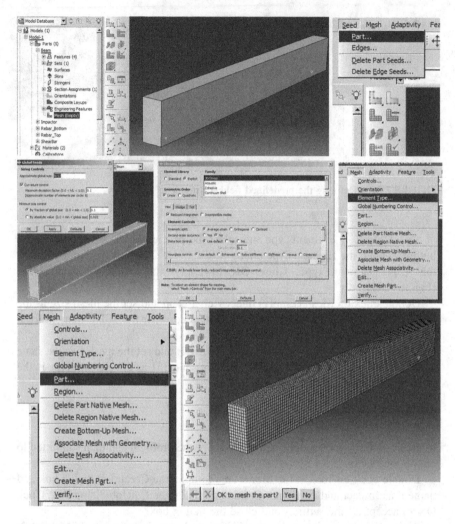

**FIGURE 7.38** Meshing beam.

Then click on the ContoursDeformedShape button to display Von Mises stress distribution on the beam, which is at most 18.7 MPa (Figure 7.44).

Replace four longitudinal bars:

Because the longitudinal bars only provide the axial stress component, from the list of variables, Select S11. As shown in the legend, the longitudinal stress in the bars is equal to 94 MPa.

To better display longitudinal bars and their stress along with section sizes, respectively, select View → ODB Display Options and check the Render Beam Profile. Then click OK to display the overview with the bar sections (Figure 7.45).

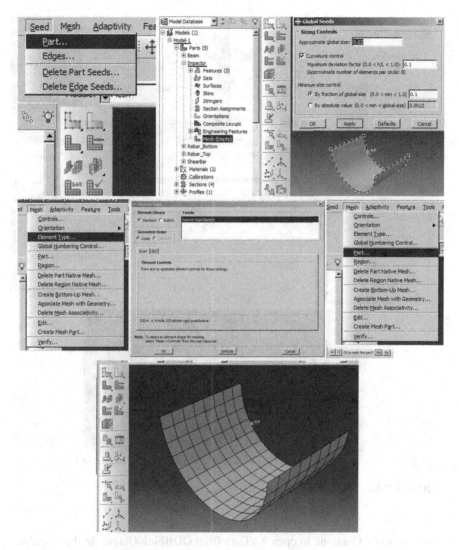

**FIGURE 7.39**  Meshing impactor.

To display all instances, click Replace all (Figure 7.46).

In the example, an area of concrete is damaged by the impact on the impactor. As only compressive damages were considered, the region could be displayed by the DAMAGEC variable. The compressive damage could be up to 0.96, as input material properties.

Select DAMAGEC from the list of variables (Figure 7.47).

To plot the beam midpoint displacements during the time, Double click on the XYData in the ResultsTree and select ODBFieldOutput from the CreateXYData

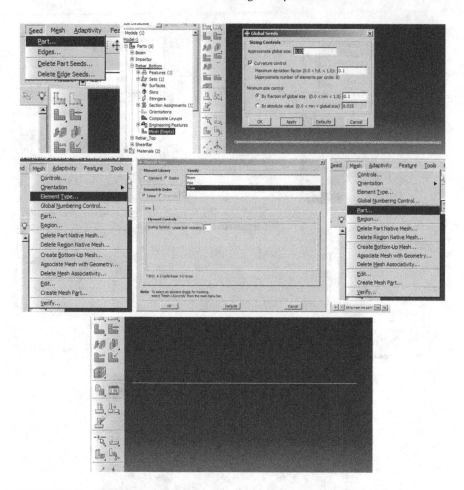

**FIGURE 7.40** Meshing rebar_bottom.

box and click Continue to open XYData from ODBFieldOutput. In the Variable tabbed page, select UniqueNodal from the Position since the displacements are recorded at nodes. Select U2 for displacement in the Y direction. Select the Elements/Notes tabbed page and click EditSelection, and then choose the node placed in the middle of the upper beam face and click Done in the prompt area to accept the selection. To draw a plot, click the Plot button. As shown in Figure 7.48, the maximum displacement is about 0.75 mm.

Select the PEEQ from the list of variables to extract the equivalent plastic strain contour (Figure 7.49).

The highest strain of plastic is at the middle of the beam at the point of impact of the impactor, which is measured at 0.012.

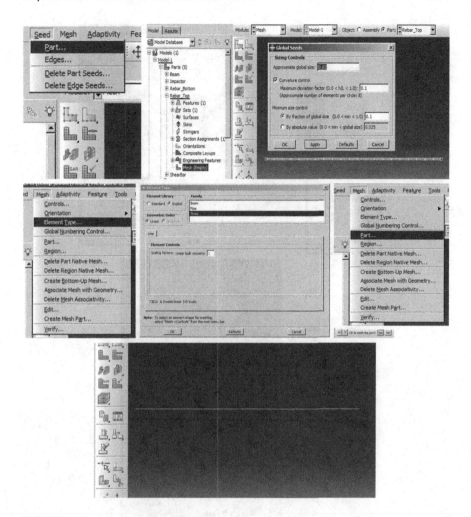

**FIGURE 7.41** Meshing Rebar_Top.

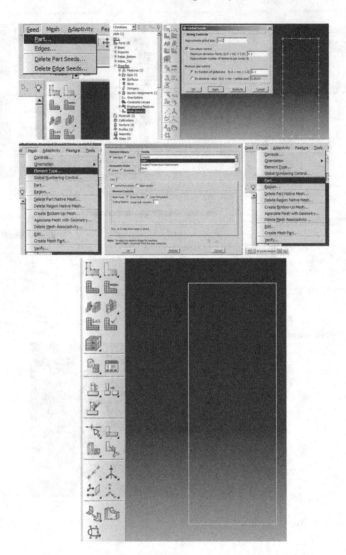

**FIGURE 7.42**  Meshing shear bar.

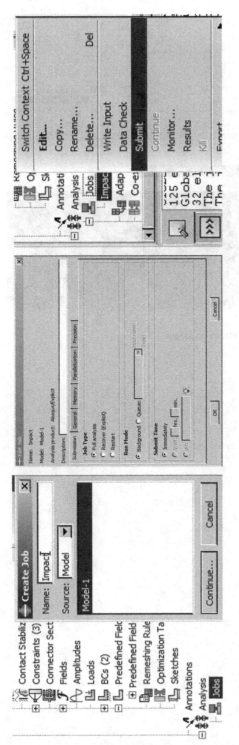

**FIGURE 7.43** Define and submit the job.

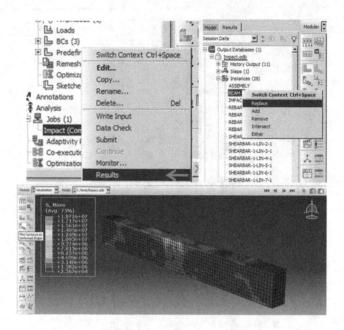

**FIGURE 7.44**   Von Mises stress on beam.

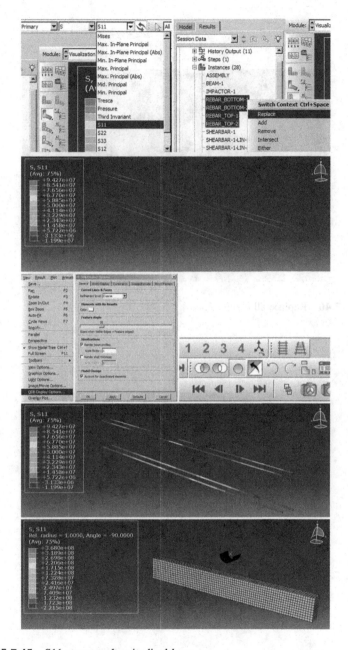

**FIGURE 7.45** S11 stress on longitudinal bars.

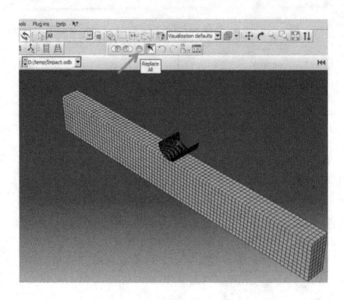

**FIGURE 7.46**  Replace all components.

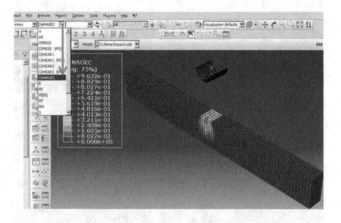

**FIGURE 7.47**  DAMAGEC on concrete.

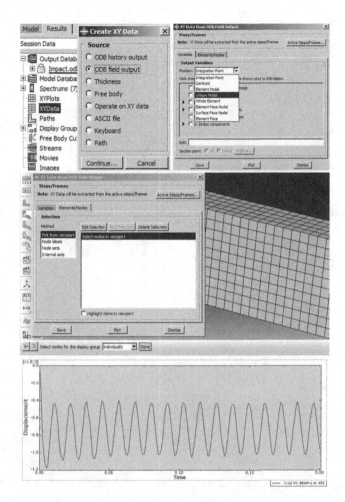

**FIGURE 7.48** Displacement time history.

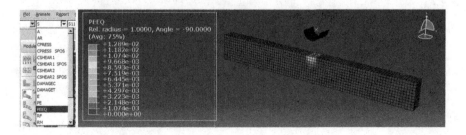

**FIGURE 7.49** PEEQ contour plot.

# 8 Eulerian Buckling of Column under Axial Load

## 8.1 INTRODUCTION

Buckling normally occurs in structural members under compression load, before reaching to the maximum compressive strength and before failure under the influence of compression stress. If the column is tall and slender, the early yielding is bound to occur due to the buckling failure. Then, buckling is a type of structure instability condition that leads to failure.

As mentioned before, buckling occurs in the structural member under pressure, which is defined as a sudden lateral deflection in a member.

When the applied load on structural member such as column reaches to the critical load, which is called the Euler's load, the column is said to enter the buckling range. In this condition, by applying any more load, the column suddenly will jump to a new state and is therefore said to be buckled. The Euler equation is given in (8.1) as:

$$P_{cr} = \frac{\pi^2 \times EI}{\left(L_{eq}\right)^2} \tag{8.1}$$

where $F$ is the most load or critical load (vertical load in the column), $E$ is Young's modulus, $I$ is representative of the moment of inertia for the column cross section, and $L_{eq}$ is the equivalent length of the column. Buckling usually occurs before the axial compressive yield stress.

## 8.2 PROBLEM DESCRIPTION

In this problem, buckling of a two-ended clamped steel column under an axial compression is investigated by aide of finite element method, and its result is checked by (8.1). Note, that $L_{eq}$ for the tow-ended clamped column is defined as $L_{eq} = 0.5 \times L$. The column's dimensions are shown in Figure 8.1.

The material properties for steel are defined as 200 GPa Young's modulus and 0.3 Poisson's ratio.

DOI: 10.1201/9781003213369-8

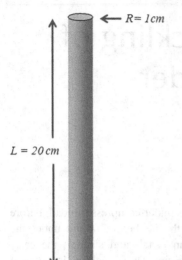

**FIGURE 8.1**   Problem dimensions.

## 8.3   OBJECTIVES

1. To simulate Eulerian buckling in the steel column
2. To investigate about buckling eigen solvers.

## 8.4   MODELING

Run Abaqus/CAE from the start menu and close the StartSession dialog box (Figure 8.2).

### 8.4.1   PART MODULE

At first, the geometry is needed to be defined. Considering that the cross section of the column is less than its length, the assumption of Beam elements is valid. So, it should be drawn as Wire.

Double click on Parts in the ModelTree to open the CreatePart dialog box. Name the part as Column and choose Wire as the Shape and consider the Approximate size as 10 and click Continue to open the Sketcher.

Select Create Lines: Connected and draw a vertical line. And select AddDimension and determine the line length as 0.2.

Click the middle mouse button twice to exit the Sketcher, as shown in Figure 8.3.

### 8.4.2   MATERIAL PROPERTIES

The properties of the material, which in this example simply consists of Young's modulus and Poisson's ratio, need to be entered in the program.

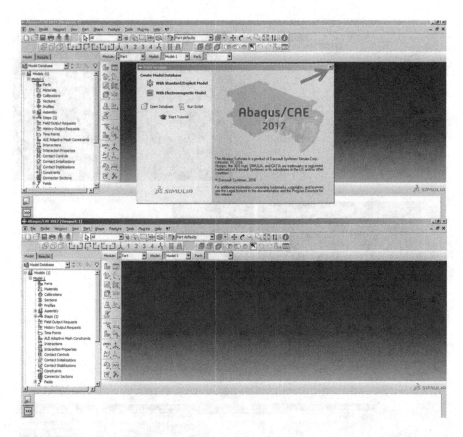

**FIGURE 8.2** Abaqus/CAE.

Double click on Materials to open the EditMaterial dialog box and enter the properties, as shown in Figure 8.4.

### 8.4.3 SECTION PROPERTIES

In the next step, it is necessary to define the section. Material properties and profile play the main role to calculate the stiffness matrix and should be defined at this point.

As the column is considered to be wired shape, it is necessary to define the beam section. Beam sections are provided as two types: Truss and Beam. The Truss can just support the axial forces, while the Beam can support all kinds of forces. As in the example, the column is clamped in two ends and, as such, bending moments will occur, So, the section should be Beam/Beam. On the other hand, a circular profile must be defined and oriented so that the software can calculate inertia.

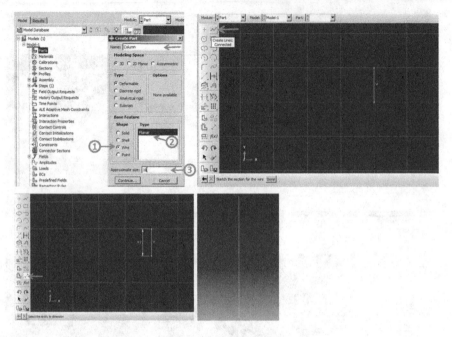

**FIGURE 8.3** Creating a part.

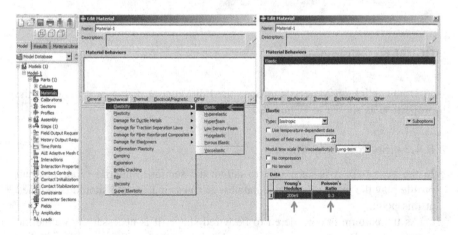

**FIGURE 8.4** Defining the material properties.

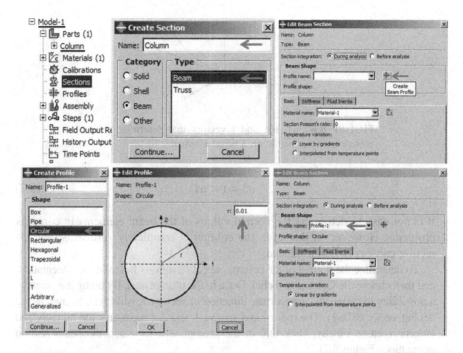

**FIGURE 8.5**   Defining the section.

Double click on Sections again and name the section as Column. Select its Category as Beam and its Type as Beam. Then click Continue to open the EditSection dialog box.

To create the profile, click CreateBeamProfile to open its dialog box. Select Circular as the Shape and click Continue. Then enter 0.01 for Radius and click OK to define the profile and close the dialog box. Circular will be shown in the beam shape box in the EditSection dialog box.

Click OK to define the section and close the dialog box (Figure 8.5).

### 8.4.3.1 Orientation
**Beam Normals:** The theory of beam sections is to calculate the exact stiffness of each element that must calculate the moment of inertia.

To determine the angle and position of the beam section profile along its length on each link, the software requires the definition of a three-axis coordinate system consisting of two vertical vectors and a tangent vector on the link. The two vectors are perpendicular to the links and are referred to as n1 and n2 local vectors, known as beam section normal, and tangential vector on the link with t, which follows the law of the right hand, according to Figure 8.6.

Among other things, the direction of vector n1 is determined by the user. Abaqus will determine the t vector, and the n2 vector is obtained by the right-hand law as expressed in Equation (8.2).

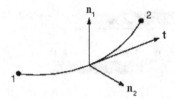

**FIGURE 8.6**  n1, n2, and t vectors that determine the orientation of the profile on a beam element.

$$n2 = t \times n1 \tag{8.2}$$

It is not easy to determine the normal values of the beam elements in complex problems. However, in this case, the column is completely vertical, so the direction can easily be introduced.

To determine the normal of the column, select Assign → BeamSectionOrientation, and then choose the Column and click Done in the prompt area. By doing this, there is a possibility to define the approximate direction of vector n1, which can be used as the default vector (0,0,-1). The red-colored arrows 1 and 2, t represents the normals and tangent vectors. Finally, click OK in the prompt area to assign the beam section orientation (Figure 8.7).

### 8.4.4  SECTIONS ASSIGNMENT

It is necessary to assign the section for the column.

Double click on SectionAssignments under the Column and select all the geometry. Then, click Done in the prompt area to open the EditSectionAssignment box and click OK to assign the section to the part and close the dialog box (Figure 8.8).

### 8.4.5  ASSEMBLY MODULE

In the next step, the assembly should be defined. Although there is one part, the assembly should be defined, as well as all degrees of freedom should be defined in the module.

Double click on Instances under Assembly to open the CreateInstance dialog box. Select Column and click OK to add it into the assembly and close the dialog box (Figure 8.9).

### 8.4.6  STEP MODULE

In the next step, the procedure type should be defined. Abaqus offers two methods for eigenvalue extraction in buckling analysis: Lanczos and Subspace.

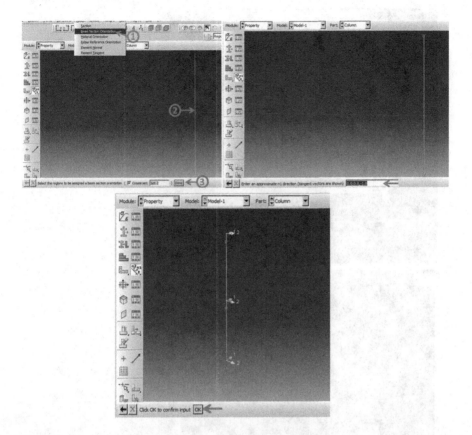

**FIGURE 8.7** Assigning an orientation.

The Lanczos method is more suitable and cost-effective when a large number of buckling eigenmodes is requested, especially for a system having many degrees of freedom. However, the Subspace method is faster and more efficient for extracting a few buckling eigenmodes.

For both methods, the user should determine the number of eigenvalues requested. Abaqus will choose a suitable number of vectors for the subspace method or suitable block size for the Lanczos method.

Note that the most important eigenvalue and its corresponding mode shape is the first one. This is the minimum critical force and most unstable mode shape, which is more likely to occur. However, for the goals of this tutorial of extracting 10 eigenvalues are intended. So, both methods can be used. However, the subspace method will be chosen. In the method, after specifying 10 for the number of eigenvalues, Abaqus will consider 18 vectors in extracting the number of eigenvalues and the corresponding mode shape.

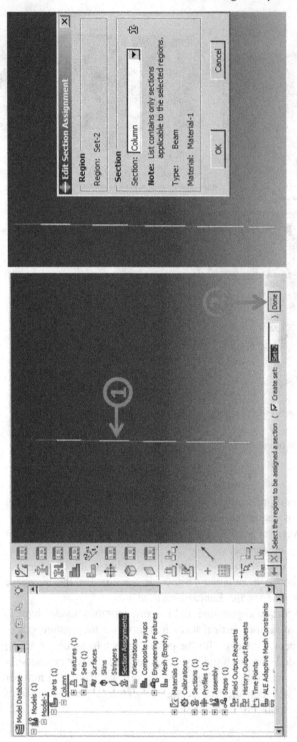

FIGURE 8.8   Assigning the section.

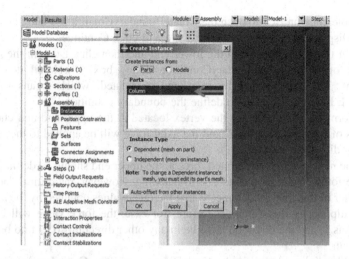

**FIGURE 8.9** Creating a column instance.

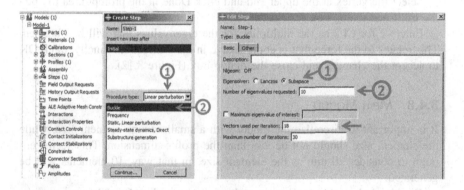

**FIGURE 8.10** Defining the analysis step.

Double click on Steps in the ModelTree to open the CreateStep dialog box. From the LinearPerturbation procedure, select Buckle and then click Continue

Chose Subspace as the Eigen solver and consider 10 for the Number of eigenvalues requested. The Vectors used per iteration will be updated to 18 that is sufficient for the problem. Leave the other parameters unchanged and click OK to define the analysis step (Figure 8.10).

### 8.4.7 LOAD MODULE

In this example, the column is clamped at the two ends, while carrying an axial force that caused buckling.

Double click on BCs to open the CreateBoundaryConditions dialog box. Then select Displacement/Rotation as the Type and click Continue.

Select the vertex located at the lower end and then click Done in the prompt area to open the EditBoundaryCondition dialog box. Check all degrees of freedom. The ends cannot be translated or rotated, which means they are clamped. Finally, click OK to define the boundary condition.

Perform the same step for the vertex located at the upper end and check all degrees of freedom except U2, and the axial force will be applied to the end and in the y direction (Figure 8.11).

In the next step, a concentrated force at the upper end should be defined. Note that the force cannot be caused due to buckling. However, the critical force is a multiplier of this load, which means that the eigenvalues obtained by the analysis are multiplied by this value. Therefore, for ease, the force value will be considered as 1, and the direction must be in any other direction so that no buckling will occur.

Double click on Loads in the ModelTree to open the CreateLoad dialog box. Chose Concentrated force as the Type and click Continue.

Select the vertex at the upper end and click Done in the prompt area to open the EditLoad dialog box.

Enter -1 for CF2 as the multiplier for the eigenvalues that will be obtained. This relates to the force that is compressive in the y direction. Finally, click OK to accept the changes and close the dialog box (Figure 8.12).

## 8.4.8 Mesh Module

It is important to discretize the model with a small enough element size, while the element size should not be less than the profile dimension. This seems enough to consider 20 mm as the element size. In that way, 10 elements will be generated.

Double click on Mesh (Empty) under Column in the ModelTree to open the Mesh module.

Select Seed → Part to open the GlobalSeeds dialog box and enter 0.02 for ApproximateGlobalSize and click OK to define the seed and close the dialog box.

Select Mesh → Part and click Yes in the prompt area to mesh the part by 10 elements (Figure 8.13).

## 8.5 ANALYSIS: JOB MODULE

In the next section, a job should be defined and submitted.

Double click on Jobs in the ModelTree to open the CreateJob dialog box. Name the job as EulerianBuckling, and then click Continue to open the EditJob dialog box. Leave the parameters unchanged and click OK to define the job and close the dialog box.

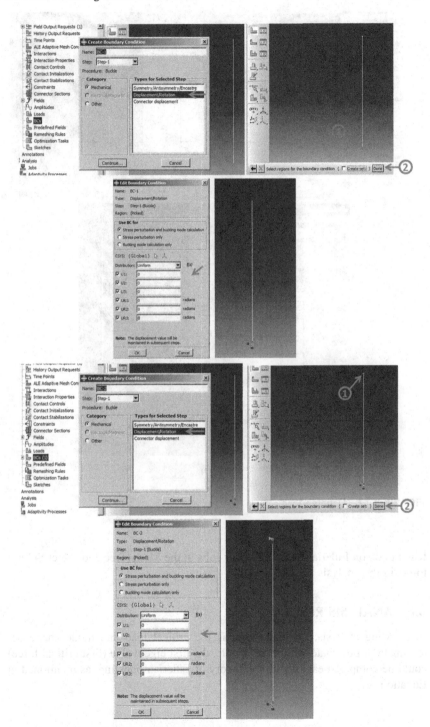

**FIGURE 8.11** Defining boundary conditions.

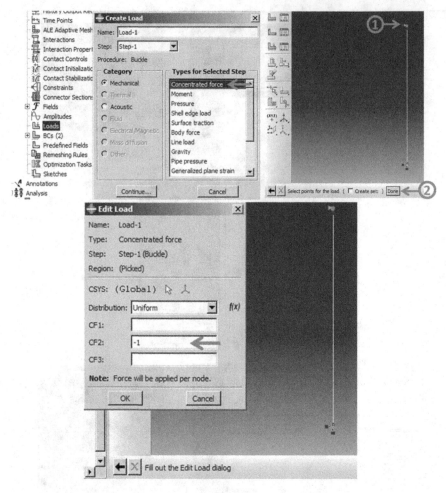

**FIGURE 8.12** Defining load.

Right click on EulerianBuckling under Jobs in the ModelTree and select Submit to begin the analysis (Figure 8.14).

## 8.6 ANALYSIS RESULTS

In buckling analysis, the main goal is the critical load and mode shape determination. Fortunately, in the example, the first eigenvalue (first critical force) could be compared easily with the theory of Eulerian buckling, as mentioned in Equation 8.1.

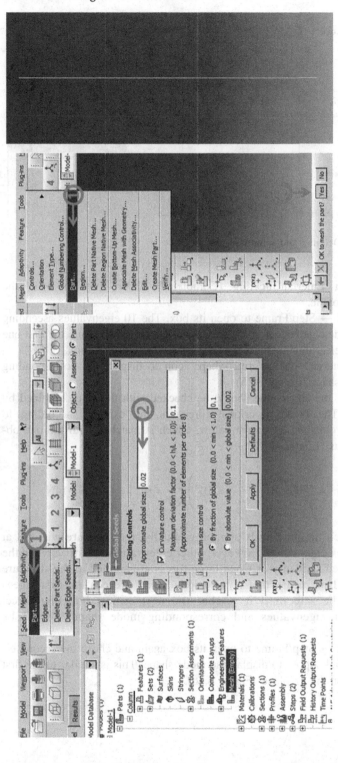

**FIGURE 8.13** Meshing.

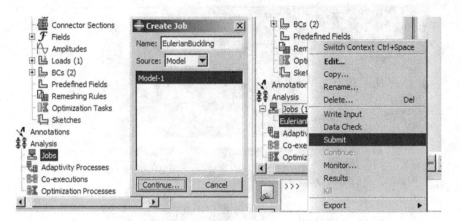

**FIGURE 8.14**  Defining and submitting the job.

Right click on EulerianBuckling (Completed) under Jobs to open the
Visualization module.

Select Result → Step/Frame to open its box. The 10 eigenvalues ascending
are extracted. The eigenvalues show the critical buckling force, and the first one
is 1.55064e6. Select this eigenvalue and click OK to close the box.

Click PlotContours on DeformedShape icon to display the corresponding
buckling mode shape (Figure 8.15).

At this stage, the eigenvalue should be checked with the value obtained by
Equation 8.1. The buckling critical force for the example according to
Equation 8.1 is obtained as shown below, which is matched with the one ob-
tained by Abaqus (1,550,640 N).

$$P_{cr} = \left(\pi^2 * 200 * 10^9 * \frac{\pi}{4} * 0.01^4\right) \Big/ \left(\frac{1}{2} * 0.2\right)^2 = 1{,}550{,}314 \text{N}$$

Usually, when the loading is applied gradually, the structure will fail at
the first buckling eigenvalue. However, other eigenvalues will arise when the
force is increased suddenly or when large forces are applied to the structure
instantaneous. This means that if the applied load exceeds the first eigenvalue, it may be safe until it reaches the next critical buckling force.
However, other eigenvalues and corresponding mode shapes should be
checked.

Select Result → Step/Frame to open its box again and chose the second eigenvalue and click Apply to display the mode shape. This is similar to the first
eigenvalue but in the yz plane (Figure 8.16).

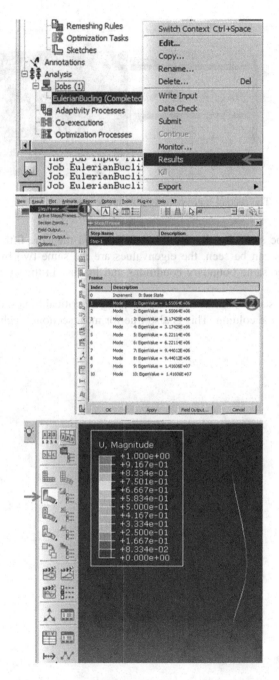

**FIGURE 8.15** The first buckling eigenvalue and the corresponding mode shape.

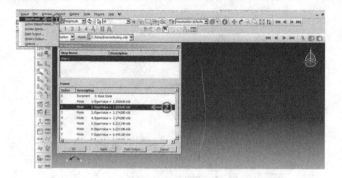

**FIGURE 8.16**   The second eigenvalue and its mode shape.

Perform the same for the third and fourth, and then the fifth and sixth eigenvalues. As can be seen, the eigenvalues are the same two by two. This is because of the same boundary conditions and loading in the xy and yz planes (Figure 8.17).

This figure shows that the more the buckling by critical forces, the more folds are created in the column. This is also true for most sections, such as boxes and pipes.

**FIGURE 8.17** Third, fourth, fifth, and sixth eigenvalues and their mode shapes.

# 9 Post-Buckling Analysis of Steel Plate Shear Wall

## 9.1 INTRODUCTION

Buckling is a bifurcation problem and the structural behavior after buckling is known as post-buckling behaviour. Post-buckling analysis is the analysis of deformation after occurring of the buckling. This is a kind of nonlinear analysis for the unstable condition after buckling and occurs in a very short amount of time with high deformation. Figure 9.1 shows three possible post-buckling solutions.

For linear buckling, only the critical load is defined. In many cases, ensuring safety against reaching the critical buckling load is sufficient. However, for some conditions, the progressive deformation of the structure during and after buckling is required. To perform a post-buckling analysis, the structure should be loaded incrementally after the structure has buckled. The analysis includes deformation after buckling called post-buckling analysis.

## 9.2 PROBLEM DESCRIPTION

In this chapter, a simple two-story frame is considered which contained a steel plate shear wall at both levels as shown in Figure 9.2, and sections of structural members are shown in Figure 9.3. The frame is exposed to a uniform and incremental load produced by a hydraulic actuator in the left end of the top beam. Due to pushing of the actuator, the frame is deflected, and the steel plate (shear wall) is buckled. So, the main goal in this example is investigating buckling and post-buckling of steel plate under applied loads.

All components are fully welded on the joints, and the lower edges of the columns and the lower edge of the lower plate are cantilevered. Also, the beams are restricted to move in the plane, and the top beam is prevented from moving vertically. The structural components are made of steel material as their mechanical properties are displayed in Table 9.1.

To conduct post-buckling analysis of the shear wall two analysis steps have been conducted. In the first step, an eigenvalue buckling analysis is performed with Abaqus/Standard on the *Perfect* structure to establish the probable collapse modes and write the eigenmodes of the default global system in the results file as nodal data.

DOI: 10.1201/9781003213369-9

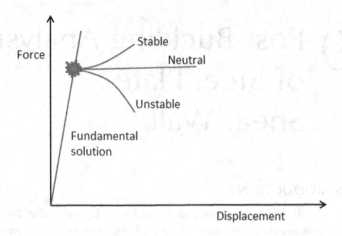

**FIGURE 9.1** Post-buckling solutions.

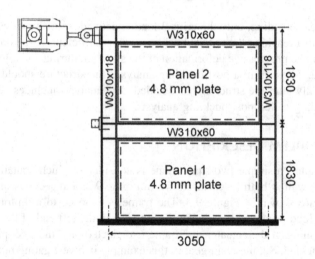

**FIGURE 9.2** Steel plate shear wall dimensions.

In the second, the Abaqus/Standard is implemented to introduce an imperfection in the geometry by adding buckling mode to the *perfect* geometry. The lowest buckling modes are often assumed to provide the most critical imperfections. Usually, these are scaled and added to the perfect geometry to create the perturbed mesh.

These steps should be defined in separate models. Therefore, the considered structure should be simulated through two models.

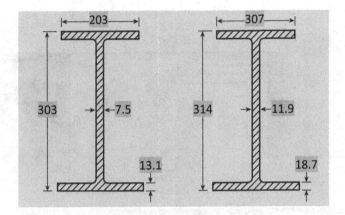

**FIGURE 9.3**    (a) W 310 × 60 mm and (b) W 310 × 118 in mm.

**TABLE 9.1**
**Structural steel mechanical properties**

| Property | Density (kg/m³) | Young's modulus (GPa) | Poisson's ratio | Plasticity | |
|----------|-----------------|------------------------|-----------------|------------|------|
| Value | 7,800 | 200 | 0.3 | Yield stress (MPa) | Plastic strain |
| | | | | 275 | 0 |
| | | | | 370 | 0.1 |

## 9.3  OBJECTIVES

1. To implement of buckling modes to subsequent analysis
2. To investigating post-buckling analysis
3. To study imperfection modeling.

## 9.4  MODELING

### 9.4.1  PART MODULE

Run Abaqus/CAE from the start menu and close the StartSession dialog box (Figure 9.4).

For simplicity, it is assumed that all components are welded and considered as one part. On the other hand, given that the thicknesses of all components are much less than the other dimensions, shell elements are more efficient to use as

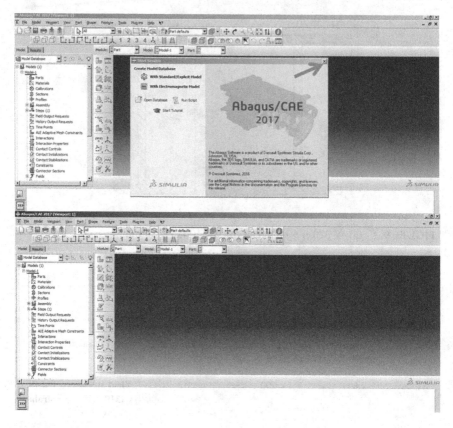

**FIGURE 9.4**    Abaqus/CAE.

they provide nodes just in one layer and are more suitable to calculate buckling mode shapes.

First, the columns should be drawn, then beams will be added to the structure, and finally shear walls should be defined.

Double click on parts in the ModelTree and open the CreatePart dialog box. Name the part as ShearWall and select Shell as the Shape and Extrusion as the Type.

Set the Approximate size as 10 and click Continue to activate Sketcher.

Select Create Lines: Connected and draw an arbitrary/section. Try this with five lines.

Choose AddConstraint and make flanges of Equal length.

Click AddDimension and set the flanges and web length, as illustrated in Figure 9.3(b).

Select HorizontalConstructionLine and draw a horizontal line in the bottom of the sketch. Set the distance between the lower flange and the line to 1.368 using AddDimension.

Click Mirror and select Copy from the prompt area. Then choose the horizontal line as a mirror line and then select all five lines as mirror entities. Finally, click the middle mouse button to perform the mirror.

Consider 3.66 as the Depth in the EditBaseExtrusion dialog box and click OK to create the part and close the dialog box (Figure 9.5).

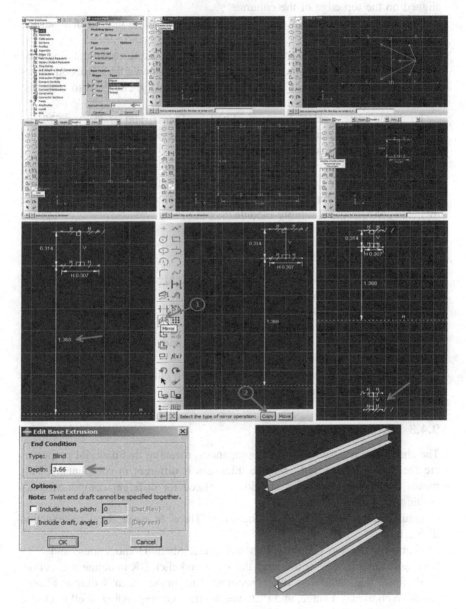

**FIGURE 9.5**  Design columns.

To define the beams, select Shape → Shell → Extrude, and choose one of the outer column flanges as the sketch plane and one of the right column edges as the sketch reference edge to activate sketcher once again.

Using Create Lines: Connected draw the *I* section and set the flanges and web length using AddDimension. Note that the upper flange of the beam should be aligned on the top edge of the column.

Perform the same for the lower beam and draw a section, and then set the dimensions.

Double click the middle mouse button to open the EditExtrusion dialog box. Consider 3.364 as the Depth. Note the red arrow; it shows the extrusion direction. Use Flip if the opposite direction is needed. Finally, click OK to close the dialog box (Figure 9.6).

To define plates, select Shape → Shell → Planar, and choose one of the column webs as the sketch plane and one of the upper horizontal edges as the sketch reference edge to activate sketcher again.

Using Create Lines: [Rectangle] draw a rectangle so that its two corners are aligned on two inner corners of the frame. Double click on the middle mouse button to finalize the part definition (Figure 9.7). Note that the shear wall will be rotated by 90° in the assembly module.

### 9.4.2 MATERIAL PROPERTIES

In the next step, material data should be defined. Mechanical properties are already determined, as shown in Table 9.1.

Double click on Materials in the ModelTree and SelectGeneral → Density and enter 7,800 as MassDensity.

Select Mechanical → Elasticity → Elastic, and enter 200e9 for Young's modulus and 0.3 for Poisson's ratio.

Select Mechanical → Plasticity → Plastic, and enter YieldStresses and corresponding PlasticStrains, as shown in Figure 9.8.

### 9.4.3 SECTION PROPERTIES

The shear wall is assumed to be homogenous, meaning that material properties are the same in all regions, but the thickness is different in components. This means that different sections should be defined for different components, according to Figures 9.2 and 9.3.

Double click on Sections in the ModelThree to open the CreateSection dialog box.

Name the section Column_Web and select Shell, Homogenous, and click Continue. Enter 0.0119 as the Shell thickness and click OK to define the section.

Perform the same for other sections and name them Column_Flange, Beam_Web, Beam_Flange, and Plate and set their corresponding Shell thickness as 0.0187, 0.0075, 0.0131, and 0.0048, respectively. Accept the default setting, as shown in Figure 9.9.

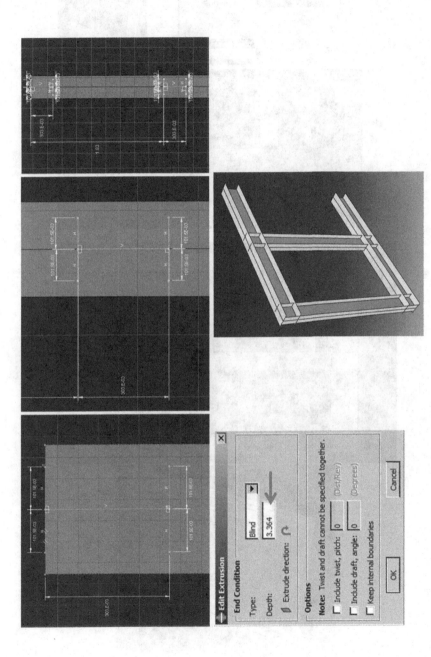

**FIGURE 9.6** Design beam.

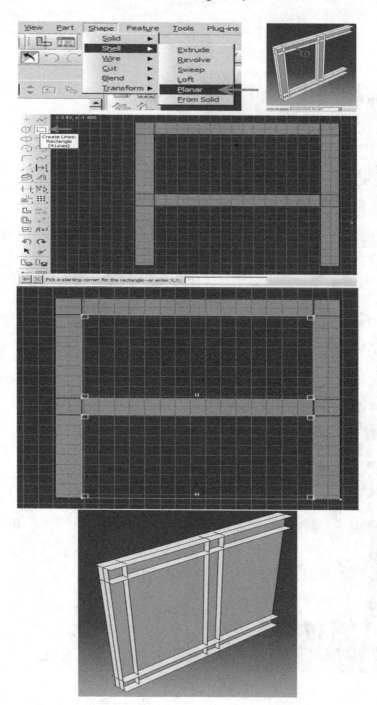

**FIGURE 9.7** Defining plates.

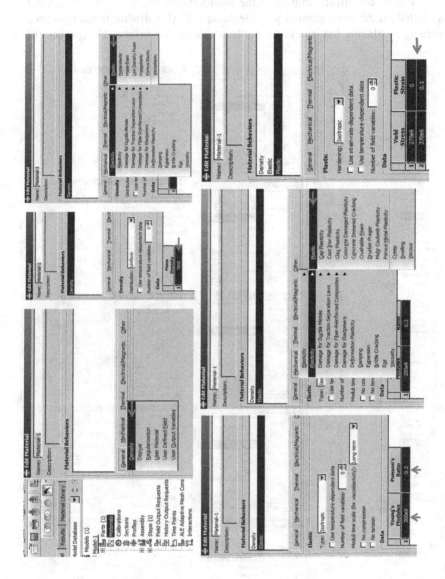

**FIGURE 9.8** Defining mechanical properties.

### 9.4.4 Section Assignment

At this point, the defined sections should be assigned to the geometry. In shell elements, nodes are placed on a surface called the reference surface. This surface is by default the middle of the shell's thickness. In other words, what is plotted on the shell geometry in the Abaqus/CAE software is the reference surface, by placing half thickness on each side of the surface. This process of

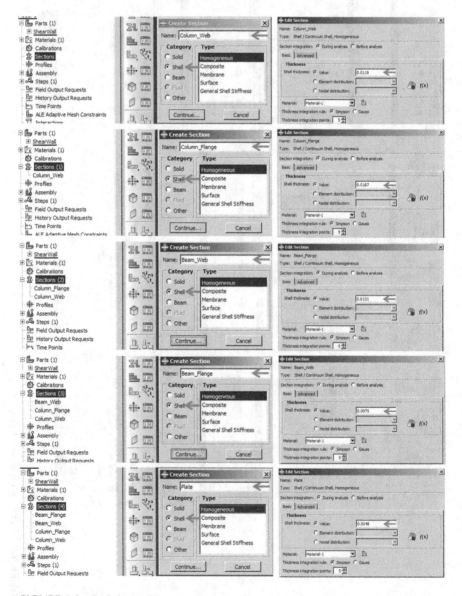

**FIGURE 9.9** Defining sections.

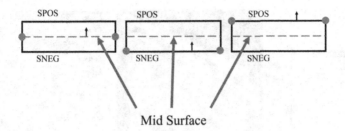

Mid Surface

**FIGURE 9.10** Reference surface positions.

extending thickness on geometry is appropriate in most cases, but in some cases, it is possible to change the surface of the surface according to the location. Figure 9.10 shows a schematic of the positions of the reference surface.

In the example, for section assignments in the column web, beam web, and shear wall, the moving reference surface is not required. However, for flanges, according to the element normal, the reference surface should be moved. So, before assigning sections to them and the moving reference surface, it is necessary to determine element normal on the flanges and then assign the section.

Double click on SectionAssignments under ShearWall in the ModelTree. Then, hold down the shift key on the keyboard and select two column webs and click the middle mouse button to open the EditSectionAssignment dialog box. Choose Column_Web and click OK to assign the section to the component and close the dialog box.

Perform the same for assigning Beam_Web and the Plate section to the associated components (Figure 9.11).

In the next step, Element normal should be determined. This is necessary for normal directions that are similar in every flange. For the convenience of modeling, flip sides so that the opposite sides of the flanges have the same color (normal) and the sides of the internal flange are brown.

Select Assign → ElementNormal to Abaqus and determine shell element normal direction by colors. Sides that are normal and outward are shown by the color brown and sides that are normal and inward are shown by the color purple. Flip normal for the sides, as shown in Figure 9.12 by holding down the shift key on the keyboard and selecting them. Then click Done in prompt area.

Finally, flange sections will be assigned. Reference surfaces should be moved according to the flange normals.

Double click on SectionAssignments under ShearWall in the ModelTree. Then hold down the shift key on the keyboard and select four column flanges and click the middle mouse button to open the EditSectionAssignment dialog box. Choose Column_Flange as the section name and Bottom surface as the Shell

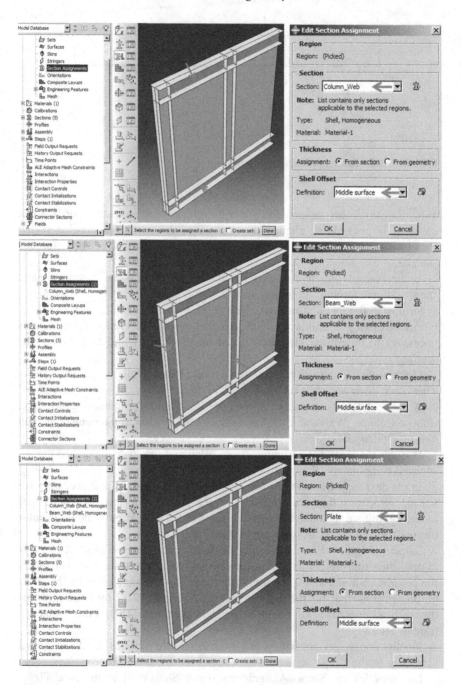

**FIGURE 9.11** Assigning section to column, beam web and plate.

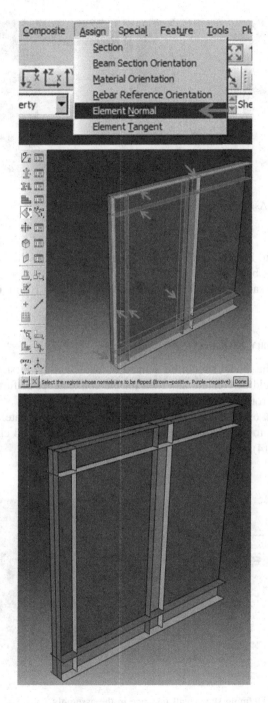

**FIGURE 9.12** Changing element normal.

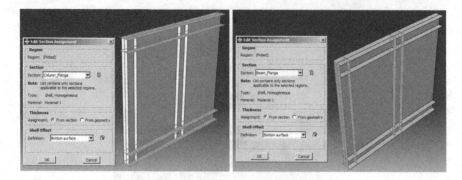

**FIGURE 9.13**   Assigning section to column and beam flange.

offset definition. Finally, click OK to assign the section to the component and close the dialog box.

Perform the same for the beam flanges by assigning a Beam_Flange section (Figure 9.13).

### 9.4.5   ASSEMBLY MODULE

At this point, the model should be defined in the assembly. First, an instance of ShearWall should be created, and then for the convenience of further modeling, it is needed to rotate as a column vertically located.

Double click on Instances under Assembly to open the CreateInstance dialog box. Click OK to add ShearWall instance in the assembly and close the dialog box (Figure 9.14).

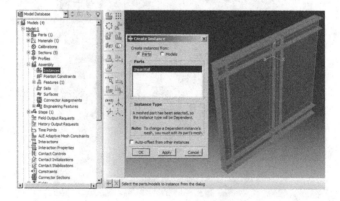

**FIGURE 9.14**   Defining shear wall instance in the assembly.

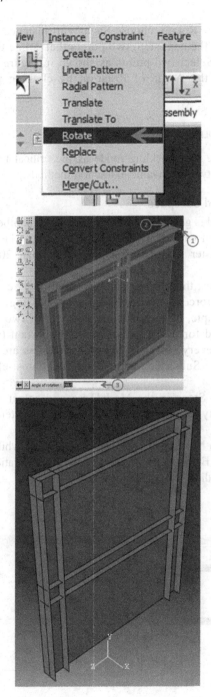

**FIGURE 9.15**   Rotating the shear wall.

Select Instance → Rotate, choose ShearWall instance and click Done in the Prompt area. In the next step, two points that define the axis of rotation should be chosen. Select two points as shown in Figure 9.15, respectively and accept 90° as the Angle of Rotation by clicking the middle mouse button.

### 9.4.6 STEP MODULE

In the next step, the first buckling mode as the critical mode should be determined. There are two eigenvalue extraction methods.

#### 9.4.6.1 Lanczos and Subspace

The Lanczos method is generally faster when a large number of eigenmodes is required for a system with many degrees of freedom. The subspace iteration method may be faster when only a few (less than 20) eigenmodes are required.

In some structures, the first, lowest buckling mode, corresponding to the minimum buckling force, is needed to run post-buckling analyses. However, in the current example, as the structure consists of two stories, the eigenmodes are separated for every story. This is important to extract the main buckling mode in every story. As four eigenmodes are enough to export buckling modes, the Subspace eigen solver is the cost-effective and better choice.

Double click on Steps in the ModelTree and select LinearPerturbation as the Procedure type and choose Buckle as the step type, then click Continue.

Consider 4 as the Number of eigenvalues requested, while Subspace chosen as the Eigen solver. Enter 20 as the Vectors used per iteration and click OK to apply and close the dialog box (Figure 9.16).

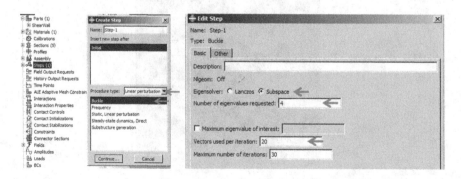

FIGURE 9.16  Defining the analysis step.

### 9.4.6.2 Coupling Constraint

Coupling constraint is necessary to define a point as the actuator support. The point should control the behavior of the corresponding part of the left column. Therefore, the left flange of the column must be partitioned, and then a reference point should be defined as the coupling control point of the flange.

Activate Part from the Module list. Select Tools → Partition to open the CreatePartition dialog box.

Choose Face as Type and Sketch as a Method of partitioning.

Hold the shift key and select the lower (left) flange as faces to partition and then click the middle mouse button.

Choose the vertical edge in the sketch to activate sketcher.

Select Create Lines: Connected and draw two lines and connect the vertices to the right and left edges. Double click on the middle mouse button to exit sketcher (Figure 9.17).

In the next step, a reference point as a coupling control point should be defined.

Activate Interaction from the Module list. Select Tools → ReferencePoint and choose the point shown in Figure 9.18.

In the next step, the coupling constraint should be defined through following steps:

Double click on Constraints in the ModelTree to open the CreateConstraint dialog. Select Coupling from the list and click Continue.

Select the reference point as the constraint control point and select Done in the prompt area.

Choose Surface as the constraint region type and hold down the shift key on the keyboard and select the corresponding faces that were generated by recent partitioning. Then, click Done in the prompt area.

Click Flip a face in the prompt area, then select the brown flange to flip it. Finally, click Purple in the prompt area to choose the faces as the constraint region and open the EditConstraint dialog box. Keep all default settings and click Ok to apply the constraint and close the dialog box (Figure 9.19).

### 9.4.7 Load Module

In the example, there are three boundary conditions. The first is about the clamping lower edges. The second is the enforced displacement at the top of the left column, and the third is about the restraint top beam and the lower beam web into the out of plan direction.

For defining the clamping boundary condition, all degrees of freedom must be eliminated from the region.

Double click on BCs in the ModelTree to open the corresponding dialog box and select Symmetry/Antisymmetry/Encastre as type and click Continue.

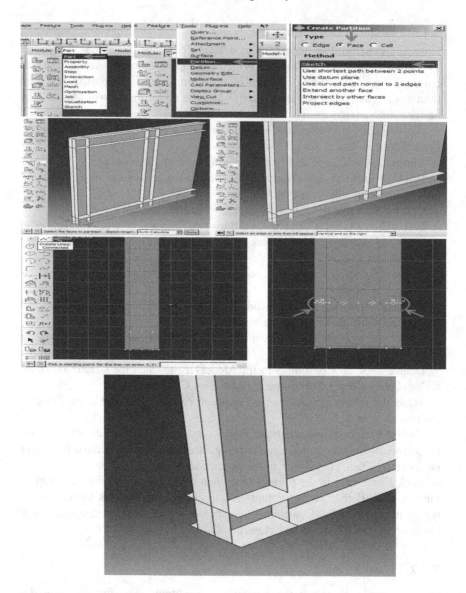

**FIGURE 9.17** Defining a partition face.

Hold the shift key on the keyboard and select all lower edges. Then click the middle mouse button to open the EditBoundaryCondition dialog box. Constraint all the DOF by toggling on ENCASTRE and click OK to apply and close (Figure 9.20).

For defining the enforced displacement boundary condition, translation along X and Z, and rotation around Y and Z should be restrained. Consider -1 as an

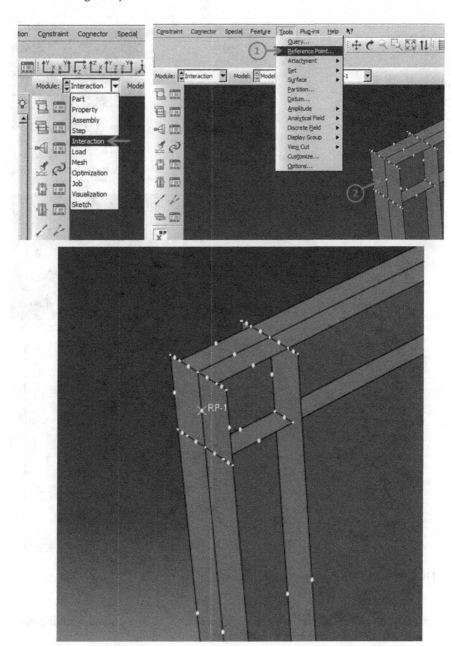

**FIGURE 9.18** Defining a reference point as a coupling control point.

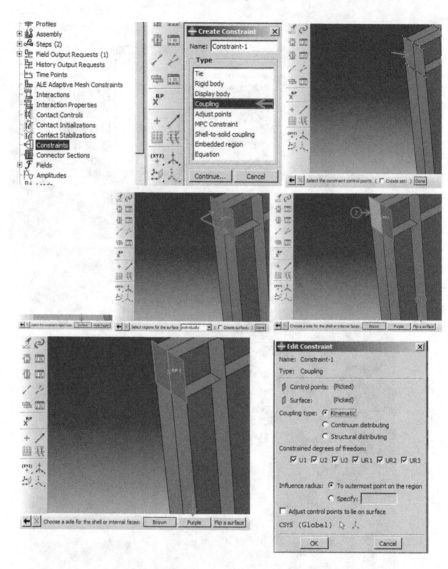

**FIGURE 9.19** Defining coupling constraint.

enforced displacement in the opposite direction of Z. The critical buckling displacement will be calculated as a multiplier of this value.

Double click on BCs in the ModelTree to open the BoundaryCondition dialog box and select Displacement/Rotation as a type of boundary condition and click Continue.

Select the reference point and click the middle mouse button to open the EditBoundaryCondition dialog box, and then fill out the box, as shown in Figure 9.21. Finally, click OK to apply the boundary condition and close the box.

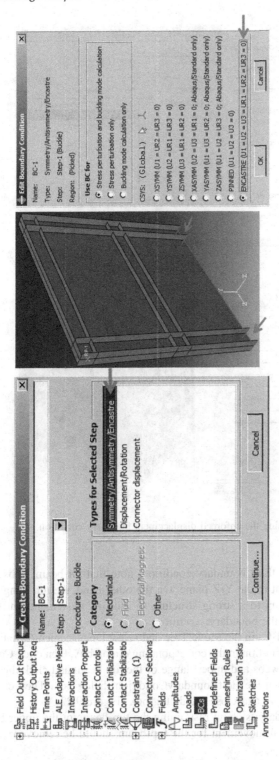

**FIGURE 9.20** Defining clamped boundary condition.

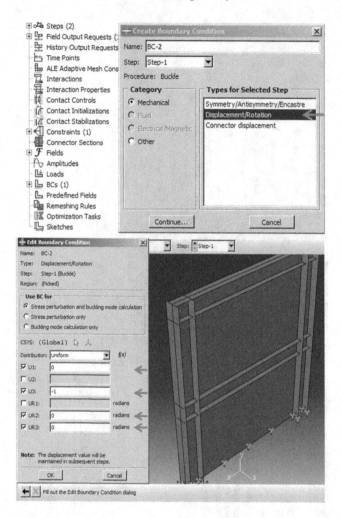

**FIGURE 9.21**  Defining the enforced displacement boundary condition.

For defining the third boundary condition, note that all beam webs are re-stricted to movement in the Y-Z plane. On the other hand, the top flange of the top beam was tied up to a strong structure to limit its movements in the Y-Z plane. For defining the boundary condition:

Double click on BCs in the ModelTree and select Displacement/Rotation as a type of boundary condition in the corresponding dialog box, then click Continue.

Select the beam webs and upper flange of the top beam and click the middle mouse button to open the EditBoundaryCondition dialog box. Then check U1 and click OK to apply and close the dialog box (Figure 9.22).

The fourth boundary is about restraining the top flange of the upper beam in the plane. So vertical movement for the top beam should be removed:

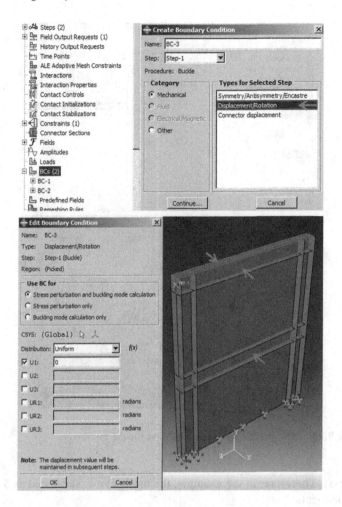

**FIGURE 9.22** Defining the third boundary condition.

Double click on BCs in the ModelTree and select Displacement/Rotation as a type of boundary condition in the corresponding dialog box, then click Continue.

Select the upper flange of the top beam by holding the shift key on the keyboard and click all portions. Finally, click the middle mouse button to open the EditBoundaryCondition dialog box, and then check U2. Click OK to apply and close the dialog box (Figure 9.23).

In the next step, meshing should be defined. This is very important in choosing a suitable element size in finite element simulations. For shells, the element size cannot be considered as small as desired. The length of the shell elements is not recommended to be less than their thickness. On the other hand, each shell should have at least two elements in any direction (except

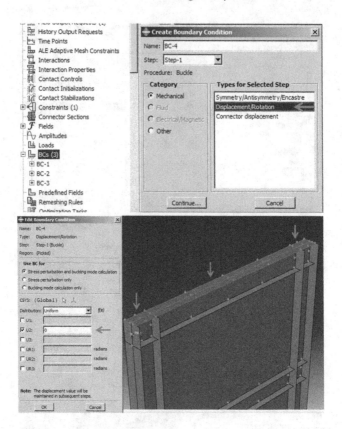

**FIGURE 9.23**   Defining the boundary condition for the top flange.

thickness). Therefore, in the example, the mesh size is suggested as 5 cm for all components.

Double click on Mesh (empty) under ShearWall in the ModelTree to activate the Mesh module.

To specify the element size, select Seed → Part, and open the GlobalSeeds dialog box. Enter 0.05 as the approximate global size and click apply to see nodes locations. Click OK to close the dialog box (Figure 9.24).

Select Mesh → Part and click Yes in the prompt area to mesh generation (Figure 9.25).

### 9.4.7.1 Writing Keywords

As eigenmodes are requested in buckling analysis, these will be used for further post-buckling analysis, which is needed to write the eigenmode in the default global system to the result file as nodal data. This could be performed by *NODEFIL*keyword. The keyword will create a *.fil file that contains the results of the eigenmodes in the global system as nodal data.

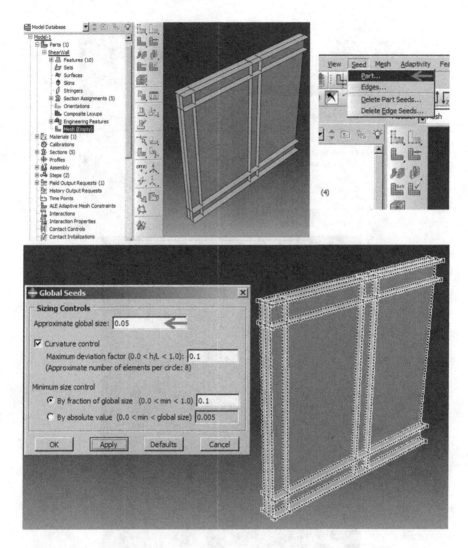

**FIGURE 9.24** Seeding.

Right click on Model-1 and select EditKeywords from the menu that appears. The EditKeywords dialog box appears containing the input file that has been generated for the model.

In the KeywordsEditor, each keyword is displayed in its own block. Only text blocks with white background can be edited. Select the text block that appears just before the *EndStep option at the last line. Click AddAfter to add an empty block of text.

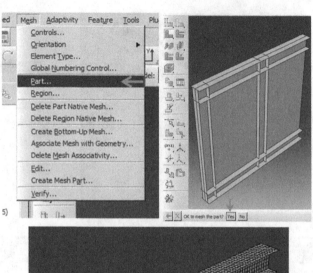

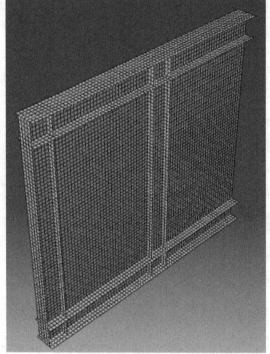

**FIGURE 9.25** Generating mesh.

In the block of text that appears, enter the following:

*\*Node file, Global = yes*

*U*

Click OK to close the keyword box (Figure 9.26).

**FIGURE 9.26** Adding a keyword.

## 9.5   ANALYSIS: JOB MODULE

This is the last step of the Part1. In this step, a job should be defined to analyze the problem and extracting the buckling mode.

Double click on Jobs in the ModelTree and name the job as Buckling. Then click Continue to open the EditJob dialog box and then click OK to create the job and close the dialog box.

Right click on Buckle underneath of Jobs in the Model tree and select Submit to begin the analysis (Figure 9.27).

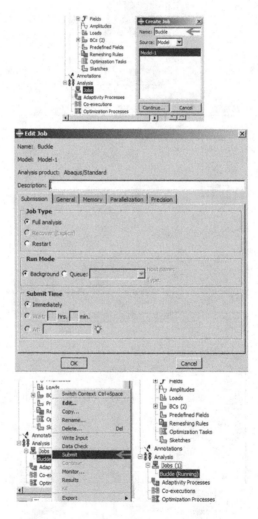

**FIGURE 9.27**   Creating and submitting a job.

## 9.6 ANALYSIS RESULTS

The main results of the buckling analysis are the mode shape and the corresponding eigenvalue. The eigenvalue in buckling analysis is a critical value that excites the problem. In the example, the model was excited by enforced displacement, so eigenvalue refers to the critical displacement that the actuator can have in order to buckle.

Right click on completed Buckle job and select Result from the menu that appears. The Visualization module will then be activated.

Select Result → Step/Frame to open its dialog box (Figure 9.28).

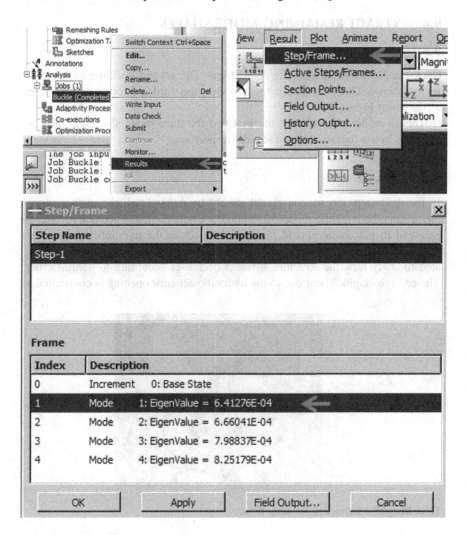

FIGURE 9.28 Extracting eigenvalue.

The calculated eigenvalue is 6.412e-4. Therefore, the critical enforce displacement could be obtained by 6.412e-4 × 1 m.

## 9.7 DISPLAY BUCKLING FOR THE FIRST BUCKLING MODE SHAPE

Click Contour plot on the deformed shape, as shown in Figure 9.29. The Legend depicts deformations relative in terms from 1.

Note that this was the first mode shape that occurred in the top story.

## 9.8 EXTRACT REMAINING MODE SHAPES

### 9.8.1 PART 1

Click the Next button from the context bar. The second buckling mode shape will be shown. Click the button again to view the third and fourth mode shapes, as shown in Figure 9.30.

This is sufficient to choose the first and the third mode shapes as they are effective in the top and bottom stories, respectively.

### 9.8.2 PART 2

In the next step, the second model should be defined, which consists of implementing the buckling mode shape into the whole complete geometry. Here, an enforced displacement should be defined to act as the hydraulic actuator and push the buckled shear wall. Finally, a load-displacement diagram will be created to study how the structure stiffness decreases according to nonlinearities. The enforce displacement due to the hydraulic actuator opening is considered as

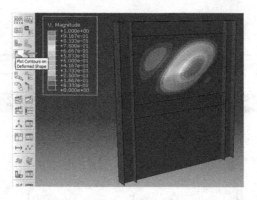

**FIGURE 9.29**   Extracting buckling mode.

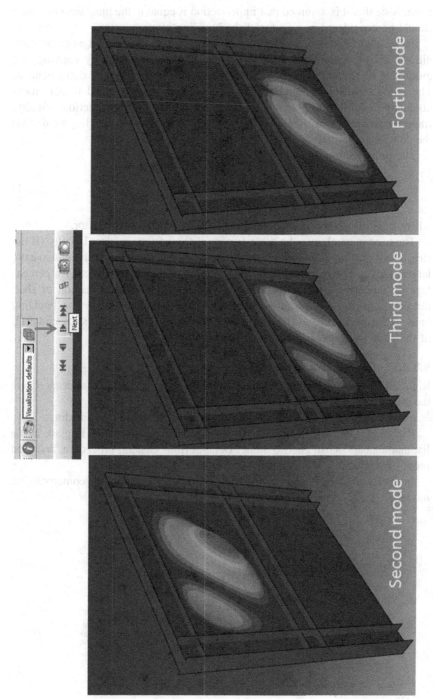

**FIGURE 9.30** The first four buckling mode shapes.

5 cm. Note that it is assumed that imperfection is equal to the plate thickness that should be considered as a factor of the buckling mode.

To simulate Part 2, it is necessary to define a new model, but as many definitions are the same as the buckling one, it could be made by copying and modifying it. Then, the buckle analysis step should be replaced by static, general. The enforced displacement analysis is nonlinear, and enforced displacement boundary condition should be modified as 0.05 in the z-direction. Finally, Imperfection keyword should be used for implementing the buckling mode into the complete geometry.

The imperfection is defined as given in Equation (9.1):

$$\Delta x_i = \sum_{i=1}^{m} w_i \varphi_i \tag{9.1}$$

where $\varphi_i$ is the $i$th mode shape and $w_i$ is the associated scale factor.

The user must choose the scale factors of the various modes. Usually (if the structure is not imperfection sensitive) the lowest buckling mode should have the largest factor. The magnitudes of the perturbations used are typically a percentage of a relative structural dimension such as a beam cross section or shell thickness. In the example, the imperfection scale factor for the first buckling mode is assumed to be 100% of the plate thickness for the upper story and 50% of the plate thickness for the lower plate.

### 9.8.2.1  To Create a New Model

Right click on Model-1 and select CopyModel from the list that appears. Copy the model as PushOver (Figure 9.31).

For replacing the analysis step to nonlinear static instead of buckling:

Right click on Step-1 under Steps in the Model tree and select Replace in the list that appears. Then choose Static, General from General procedure, and then press Continue.

In the Basic tabbed page change Nlgeom to include nonlinear geometry in the model.

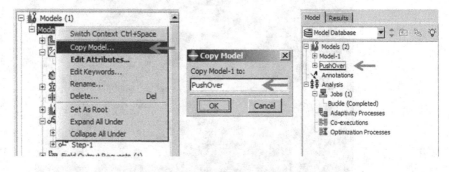

**FIGURE 9.31**  Creating a new model by copying Model-1.

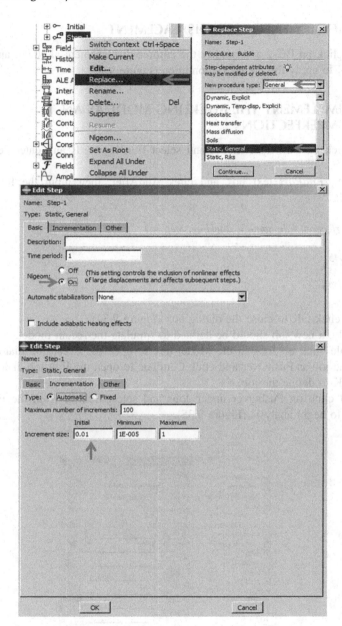

**FIGURE 9.32** Replacing the analysis step.

In the Incrementation tabbed page, set the initial increment size to 0.01. The value can ensure most nonlinear problems to converge incrementally. Then, click OK to define the analysis step and close the dialog box (Figure 9.32).

## 9.9  MODIFY ENFORCED DISPLACEMENT

Double click on BC-2 and modify the enforce displacement as -0.05, and then click OK to apply the boundary condition (Figure 9.33).

## 9.10  IMPLEMENT THE BUCKLING MODE AS AN IMPERFECTION

Right click on PushOver model and select EditKeywords and enter keyword editor.

Select *EndPart* text box and click AddAfter, then input the following keyword:

*Imperfection, File name = Buckling, Step = 1*

*1,0.0045*

*3,0.00225*

Finally, click OK to close the dialog box (Figure 9.34).

Finally, a new job should be defined to simulate the recent model.

Double click on Jobs in the Model tree to open the CreateJob dialog box. Name the job as PushOver and click Continue to open the Edit job dialog box. Click OK to define the job.

Right click on PushOver under Jobs and select Submit from the list that appears to begin analysis (Figure 9.35).

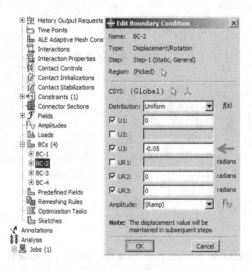

**FIGURE 9.33** Modifying the enforced displacement.

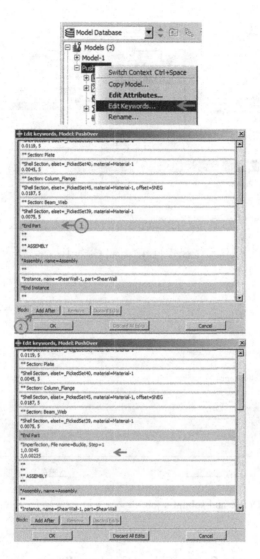

**FIGURE 9.34** Adding imperfection keyword.

## 9.11 RESULT OF PART 1

The main result is for checking plasticity regions and stress concentration, as well as a force-displacement diagram to check how structural stiffness decreases. All output can be achieved in the Visualization module.

To activate the Visualization module, right click on the PushOver completed the job, and select Result from the list that appears (Figure 9.36).

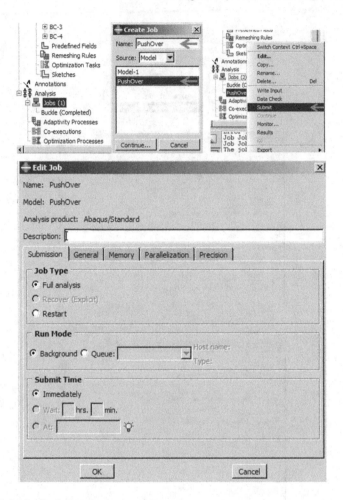

**FIGURE 9.35** Define and submit a job.

### 9.11.1 Equivalent Plastic Strain Contour Plot

Select PEEQ in the variable combo box. This implies equivalent plastic strain and is good criteria to check plasticity. As can be seen, the maximum plastic strain is about 0.0454 that occurred near the actuator (Figure 9.37).

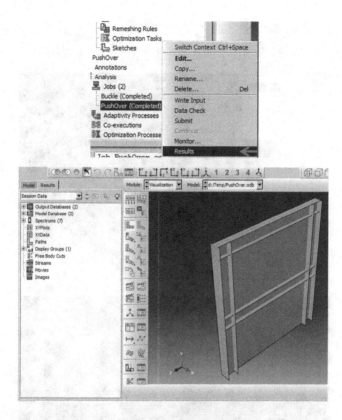

**FIGURE 9.36**   Activating the visualization module.

## 9.11.2  PLOT STRESS CONCENTRATION

Select S in the variable combo box and keep Mises as the default invariant. This implies Von Mises stress and is good criteria for checking stress in ductile metals. As shown, many areas are yielded due to the enforce displacement, especially plated that was buckled (Figure 9.38).

Next, the force-displacement diagram will be created. Note that force-time and displacement-time diagrams should be extracted separately for the reference point, and then they are combined to create a force-displacement diagram.

To create the reaction force and displacement diagrams:

Double click on XY Data in the Results tree to open the dialog box, then select ODB field output as the Source and click Continue to open its dialog box.

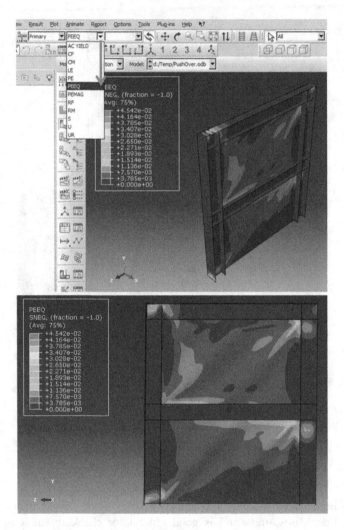

**FIGURE 9.37**  Plotting plastic strain.

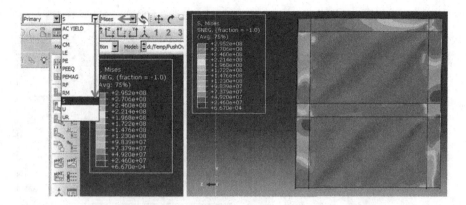

**FIGURE 9.38** Plotting Von Mises stress contour.

In the Variable tabbed page, select UniqueNodal as Position, then select RF3 and U3, which refers to the reaction force and displacement in the z-direction.

To determine the reference point, as the output position, in the Elements/Notes tabbed page, select Node sets as the Method and AssemblyConstraint-1 as a set. Then click Save and OK in the SaveXYData dialog box which implies the plot names. Finally, click Dismiss to close the dialog box (Figure 9.39).

## 9.12 COMBINE THE REACTION FORCE AND DISPLACEMENT DIAGRAMS

Double click on XY Data in the Results tree and select Operate on XY Data as a Source, then click Continue to open its dialog box.

Choose Combine as Operator and then double click on U:U3PI and RF:RF3PI data, respectively, to copy their data in combine operator.

Consider minus for both diagrams. This is because the loading on the structure given it moves in the opposite direction of the Z-axis and has produced negative values. By the minus, this means the values will be positive.

Finally, click PlotExpression to create a force-displacement diagram and click Cancel to close the dialog box (Figure 9.40).

It can be seen that stiffness decreases incrementally, especially after 1-cm movement.

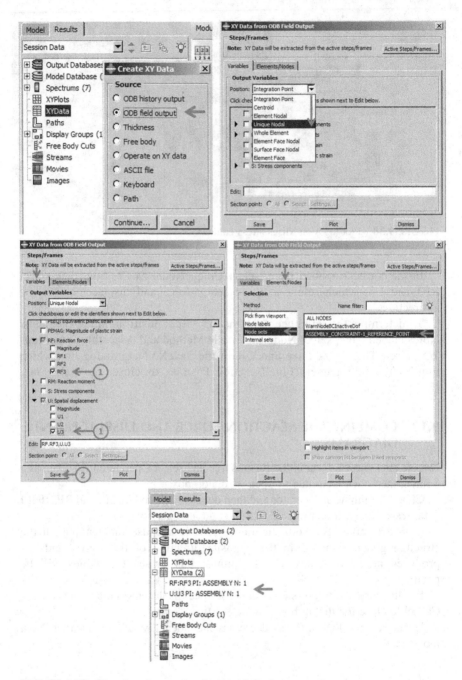

**FIGURE 9.39** Creating the reaction force and displacement diagrams.

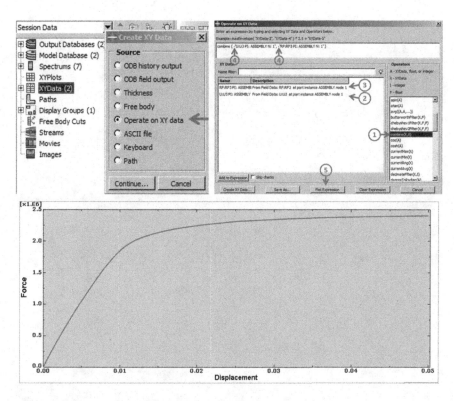

**FIGURE 9.40** Plotting the force-displacement diagram.

# Index

Printed in the United States
by Baker & Taylor Publisher Services